Introduction to Numerical Methods

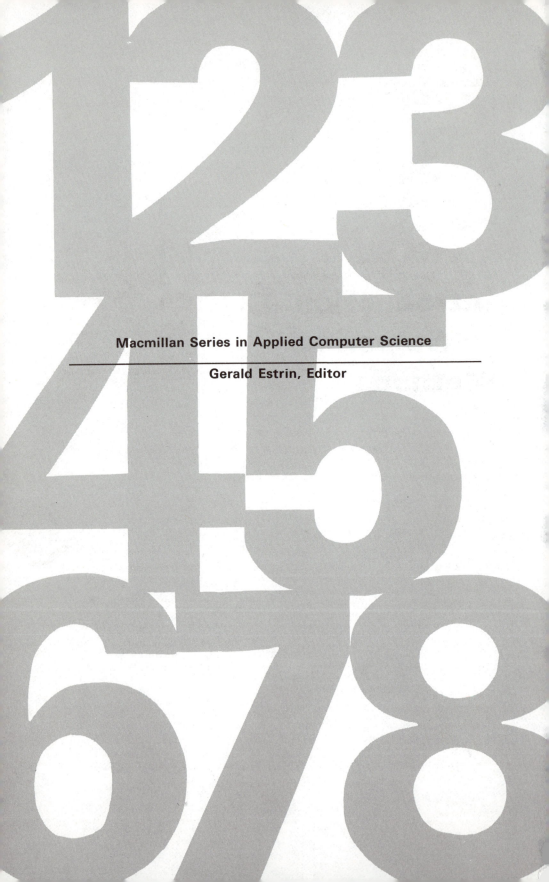

Macmillan Series in Applied Computer Science

Gerald Estrin, Editor

Peter A. Stark

Queensborough Community College
of the City University of New York

Introduction to Numerical Methods

Macmillan Publishing Co., Inc.
New York

Collier Macmillan Publishers
London

Library of Congress catalog card number: 77–85773

Macmillan Publishing Co., Inc.
866 Third Avenue, New York, New York 10022

Collier-Macmillan Canada, Ltd.

Printed in the United States of America

PRINTING   19 20 21   **YEAR**   3456

to Lois

Preface

This text is for an introductory course in what is commonly called *numerical analysis*, *numerical methods*, or even *numerical calculus*. While it parallels the development in Course B4 on Numerical Calculus in the proposed Curriculum in Computer Science issued by the Association for Computing Machinery, this book is designed for any science or engineering student who has completed his first course in calculus, and who has at least a passing knowledge of elementary computer programming in FORTRAN.

This is a practical book for the student who, in addition to seeing the theory of numerical methods, also likes to see the results; the predominant emphasis is on specific methods and computer solutions. It often points out where the theory departs from practice, and it illustrates each method of computer solution by an actual computer program and its results.

Some prior knowledge of FORTRAN programming is assumed, and the reader may easily obtain it from any one of the many excellent FORTRAN texts available. Only a very elementary knowledge is actually required, since the simple programs presented herein are amply annotated.

A short note about these sample programs is in order. They are generally written in the most straightforward manner possible, and seldom use any of the tricks which a professional programmer would employ. The emphasis here is on simplicity and clarity. Often a few extra steps, judiciously added here and there, might result in a more elegant, faster, and possibly more accurate program. Still, the purpose of this text is to teach numerical methods, not FORTRAN programming.

The student is also warned against expecting identical results if he decides to try these sample programs on another computer. Identical programs will generally yield slightly different answers on different computers, depending on the FORTRAN system in use. The sample programs herein were tested on an IBM 1130 computer in either standard or extended precision, depending on the point being illustrated.

The author would like to express his thanks to his many students who used this material in note form and made many suggestions, especially Mrs. Doris Topel and Mrs. Sylvia Neiman, and to Michael Eichorst and Anthony Maffetone for their able help in preparing these notes.

P.A.S.

Contents

3

Roots of Equations

4

Nonlinear Simultaneous Equations

5

Matrices, Determinants, and Linear Simultaneous Equations

6

Numerical Integration   *192*

7

Ordinary Differential Equations   *235*

8

Interpolation and Curve-Fitting   *273*

Introduction
to Numerical
Methods

Principles of
Computer Operation

When we think of a computer we often imagine a giant electronic "brain" which never makes an error. How wrong can we be!

Look at this simple program:

```
          SUM = 0.
          DO 10 I = 1, 10000
    10 SUM = SUM + 0.0001
          WRITE  (3, 20) SUM
    20 FORMAT (F15.8)
          CALL EXIT
          END
```

Simple, you say. Take a quantity SUM, make it equal to zero, and then add the quantity 0.0001 to it 10,000 times, and the answer is 1.00000000, exactly.

Things are not always what they seem. If we actually try the program on a typical computer, we might easily get an answer of

```
    0.99935269
```

Granted that the error may be small—only 0.00065 or so, or about 0.065 percent—still, the answer is wrong. It certainly destroys our notion of a

1

computer's infallibility, and makes us wonder about the results of other, more complicated calculations.

To see why a computer makes such errors at all (and how to go about reducing errors), we will spend this chapter on a quick view of how a computer does its arithmetic.

1.1 The Binary Number System

We humans do our arithmetic in the so-called *decimal* number system, where the number "ten" takes on a special importance; it developed quite naturally from the fact that we have ten fingers. In the computer, on the other hand, internal calculations are done in the *binary* number system, which is quite different.

The decimal number system, for example, has ten different digits, 0 through 9. The binary number system has only two different digits, 0 and 1. It does not use the digits 2 through 9 since they are not needed. As an abbreviation, the two *binary digits* are usually called *bits*.

In many other ways, however, the binary number system is similar to the decimal system. For example, in both systems the position of a particular digit in a number has a great importance. Thus in both number systems the left digits of a number are more important than the right digits since they have a greater value.

In a decimal integer, each digit has 10 times the value of the digit to its right. Thus the last digit on the right of an integer is the "units" digit, the next is the "tens" digit, the next is the "hundreds" digit, and so on. Thus, for example, the decimal number 532 can be interpreted as

$$5 \text{ hundreds}$$
$$+ 3 \text{ tens}$$
$$+ 2 \text{ units.}$$

A more concise notation expressing the same thing is

$$532 = (5 \times 10^2) + (3 \times 10^1) + (2 \times 10^0).$$

The last digit on the right has a value of 10^0 or 1, the next digit to its left has the value 10^1, the next has the value 10^2, and so on. The same principle applies to non-integer numbers such as 14.37, where the digit to the *left* of the decimal point has the value of 10^0 with the powers of 10 increasing to the left, whereas the digit to the *right* of the decimal point (the "tenths" digit) has a value of 10^{-1} with the powers becoming more negative to the right. Thus we can write

$$14.37 = (1 \times 10^1) + (4 \times 10^0) + (3 \times 10^{-1}) + (7 \times 10^{-2})$$
$$= \quad 10 \quad + \quad 4 \quad + \quad 0.3 \quad + \quad 0.07.$$

Binary numbers can be expressed in exactly the same way, except that they only contain the digits 0 and 1, and that adjacent digits differ in value by a power of 2, rather than by a power of 10. For example, the binary integer 1011 can be expressed as

$$1011 = (1 \times 2^3) + (0 \times 2^2) + (1 \times 2^1) + (1 \times 2^0).$$

Changing into decimal numbers, we see that the binary 1011 is equal to $8 + 2 + 1$, or a decimal number 11. Because each digit of a binary number can only be a 0 or a 1 and therefore can carry less information than the 10 digits of the decimal number system, binary numbers are generally much longer than their decimal equivalents. For example, the decimal number 4094 is 111111111110 in binary.

Noninteger binary numbers can be expanded into powers of 2 and converted into decimal almost as easily as integers. For example, the binary number 110.11 can be written as

$$110.11 = (1 \times 2^2) + (1 \times 2^1) + (0 \times 2^0) + (1 \times 2^{-1}) + (1 \times 2^{-2}),$$

where the exponent of 2 starts with 0 to the left of the decimal point and increases to the left, and starts with -1 to the right of the decimal point and becomes more negative to the right. In the case of the binary 110.11, we see that it equals a decimal

$$4 + 2 + \tfrac{1}{2} + \tfrac{1}{4} = 6.75.$$

A decimal integer can always be exactly represented by a binary integer, although the binary integer may be quite long, for every integer can be expressed as a sum of powers of 2. But this is not true of fractional numbers.

In general, it can easily be shown that a rational fraction can be exactly expressed with a finite number of binary digits only if it can be expressed as the quotient of two integers p/q, where q is a power of 2; that is, $q = 2^n$ for some integer n. Obviously only a small proportion of rational fractions will satisfy this requirement.

Thus even some simple decimal fractions cannot be exactly expressed in the binary number system. For example, the simple decimal fraction 0.1 is an infinitely long binary number 0.000110011... with two 1's and two 0's repeating themselves forever. Every repeating decimal (such as 0.33333333...) is also repeating in binary, but other numbers which are not repeating decimals in the decimal number system may become repeating in binary.

It is not always so easy to spot repeating fractions in the binary number system. The decimal number 0.0001 from the program at the beginning of this chapter is a repeating binary fraction which begins with 0.0000000000000110100011011011110001... and lasts for 104 bits before starting to repeat with the part ...11010001101101110001.... Every 104 bits from now on, this binary fraction starts to repeat itself.

We are now on the track of a major source of computer errors. Obviously an infinitely long binary fraction cannot be stored or used in a digital computer, and so some finite number of bits must be used and the rest discarded. This automatically leads to a small error which, by being repeated many times, can lead to a large error in the final answer.

1.2 Integer Arithmetic Operations in FORTRAN

In general, the binary numbers used in a digital computer are limited to a maximum length by the design of the machine. This length varies from computer to computer, and is usually the result of a number of compromises. For accuracy in FORTRAN and in scientific calculations, a long length is desirable; a short length, on the other hand, is significantly cheaper and more useful in a computer used for business calculations and processing.

The fixed-length group of binary bits is generally called a computer *word*. Word lengths range from 12 bits (in the Digital Equipment Corporation PDP-8 computer to 60 bits (in the Control Data Corporation 7600 computer). In some computers a longer word (such as 32 bits in the IBM 360 computer) is broken into smaller pieces called *bytes* (8 bits each) for ease in handling.

Each computer word, whatever its length, is treated as a single number, although under certain conditions several words can be used to hold one number if more accuracy is needed. As an example, let us consider the 32-bit long word in the IBM 360 computer.

Of the 32 bits in the word, the first bit is used to hold the sign of the number; a 0 means plus, and a 1 means minus. The remaining 31 bits can hold a binary number from 0 up to 1111111111111111111111111111111. If we convert this number to a decimal integer, we obtain

$$(1 \times 2^{30}) + (1 \times 2^{29}) + (1 \times 2^{28}) + (1 \times 2^{27})$$
$$+ \cdots + (1 \times 2^{1}) + (1 \times 2^{0}),$$

which equals the decimal integer 2,147,483,647 (equal to $2^{31} - 1$). Hence each 32-bit computer word can store a number as large as 2,147,483,647, positive or negative.[1]

Note, however, that this applies only to integers on a computer with a 32-bit word length. A computer with a 16-bit word, such as the IBM 1130, has

---

[1] In addition, the number 10000000...000000, instead of meaning "−0" or "minus zero," would be defined as −2,147,483,648, giving the computer the range from +2,147,483,647 to −2,147,483,648 for 32-bit integers.

a maximum limit of $2^{15}-1$, or 32,767, and other computers may have other limits.

In order to use integer arithmetic in FORTRAN we use variable names beginning with the letters I through N, or use an INTEGER specification statement at the beginning of our program. There are many reasons for specifying integer arithmetic in a FORTRAN program, one of them being the ability to exactly store decimal integers with no error. On the other hand, integer arithmetic has the disadvantage of not being able to handle fractions, and of having a rather low limit on the size of numbers—somewhat over 2 billion in a 32-bit computer, and only 32 thousand in a 16-bit computer.

1.3 Real (Floating-Point) Arithmetic Operations in FORTRAN

Floating-point arithmetic, also sometimes called *real* arithmetic, offers a way around the limitations of integer arithmetic. In floating-point calculations we can represent both fractions and very large numbers; the price for all of this, however, is accuracy and, in some cases, speed.

As an example, let us again consider a 32-bit word length computer. We can draw a picture of a word as follows:

The first bit is called bit 0, the second bit is called bit 1, and the last bit is called bit 31. When this word is used to store a binary integer, we can draw the picture as follows:

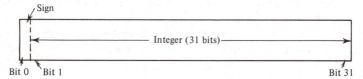

As we can see, bit 0 is used for the sign, and bits 1 through 31 are used for a 31-bit integer.

The following picture shows how the same 32-bit word is used to store a floating-point number:

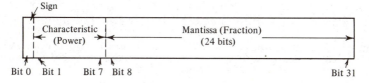

As before, bit 0 is used to store the sign of the number. But this time only the last 24 bits (bits 8 through 31) are used to actually store the number, which is stored not in its full form, but as a fraction smaller than 1. Bits 1 through 7 hold a 7-bit binary number indicating a power of 2 by which the fraction must be multiplied to give the original number.[2] The language of logarithms is applied to the fraction and power—the fraction is called the *mantissa*, and the power is called the *characteristic*.

As an example, the decimal number 1 can also be written as $\frac{1}{2} \times 2^1$; here $\frac{1}{2}$ becomes the mantissa, and 1 (the exponent of 2) becomes the characteristic:

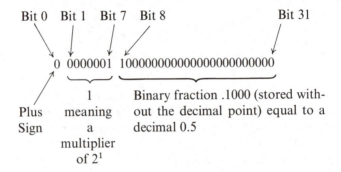

Note that the same decimal number 1 can also be written as $\frac{1}{4} \times 2^2$, or

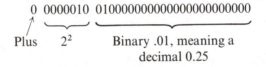

As we can see, there are many ways of storing the same number, depending on the number of initial zeroes in the binary fraction (mantissa). But if the number to be expressed has many digits, adding extra zeroes at the left of the mantissa may mean we must remove digits from the right, thereby reducing accuracy. As a result, numbers are generally stored without leading zeroes in the mantissa, and are called *normalized* when they have no such leading zeroes. After arithmetic operations such as addition or subtraction, numbers often have initial zeroes in their mantissas and the computer automatically normalizes them before saving them in memory. The object is to preserve as many significant digits of each number as possible, without saving useless zeroes.

As another example, we can see how a decimal number such as 3.5 is stored in the computer. In binary this number is 11.1, but as a floating-point number

---

[2] The power is actually a 6-digit binary number, with bit 1 used as a sign to indicate whether the power is positive or negative. In some computers a power of 8 or power of 16 is used rather than a power of 2.

it is stored as the binary fraction .111 (equivalent to a decimal 0.875) multiplied by 2^2, or 4. In terms of a binary floating-point number, it is

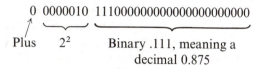

The largest number that can be stored in a 32-bit word using the floating-point notation above is

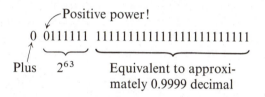

and thus the range of numbers stored in binary floating point is from about 2^{-64} (if the characteristic is negative) to almost 2^{63}; in decimal, this is a range from about 10^{-19} to about 10^{+18} in magnitude (and includes both positive and negative numbers, as well as zero). With some slight changes this range can be increased even farther, though at a slight loss of accuracy.

Obviously, floating-point notation has many advantages. Not only can it handle fractions, which cannot be handled with integer arithmetic in FORTRAN, it can even handle a much wider range of numbers. The range of $10^{\pm 18}$ or so (and $10^{\pm 63}$ in some computers) is enough to handle most practical problems we may encounter in practice.

But there are several prices to pay for this convenience. One of these is the additional time required by the computer to process floating-point numbers as compared with integers. The additional time is slight for computers with built-in electronic circuitry for floating-point calculations, but may increase the computation time several dozen times if the computer does not have such circuits and must use rather involved programs to do the same thing. –

But much more important is the loss of accuracy. In the 32-bit word we are using for illustrative purposes, integer arithmetic uses 31 bits for the actual number. Using these 31 bits we can express a number such as 2,147,483,647, *exactly.* That is, we can express a number having ten significant digits with no error. But using the same 32 bits for a floating point number leaves only 24 bits for the actual number; using these 24 bits we can now express only about seven digits of a decimal number with accuracy. The 24-bit mantissa essentially represents the first seven significant digits of the decimal number, while the 7-bit characteristic merely provides information on the exact location of the decimal point in that number.

1.4 Double-Precision Arithmetic Operations

Double-precision arithmetic is an effort to increase the accuracy of floating-point calculations by adding more bits to the mantissa. This is done by adding another computer word, and using its additional bits for the mantissa. Using the same 32-bit computer for illustration, we can now use two computer words instead of one, and add 32 bits more to the mantissa. The resulting 56 bits in the mantissa provide almost 17 decimal digits of accuracy, instead of just 7 as in normal (*single-precision*) floating-point arithmetic. Note that, despite the use of the word double, double-precision gives more than double the precision in terms of significant digits.

The disadvantage of double-precision arithmetic is the use of another computer word, increasing the memory size required by the computer program. Moreover, double-precision arithmetic is usually performed by program, rather than by circuitry, so that much more time is required to perform the program.

1.5 Roundoff Error in Floating-Point Programming

The terms *seven-digit accuracy* or the like applied to floating-point arithmetic do not really tell the story. Just having 24 bits for a mantissa and therefore expecting an answer accurate to seven decimal digits (or any other accuracy depending on the computer) is misleading; in general, the accuracy of the final answer will be much worse.

Consider the sample program at the start of this chapter. It was run on a computer which used a 24-bit mantissa in single-precision. Is our answer accurate to seven decimal places?

Definitely not; the answer of 0.9993526 is only accurate to three significant figures. To be specific, we might even say that none of the digits is accurate since the answer should be 1.0000000 and we got none of the right digits. However, since the error is only 0.0006474, we say that the first three digits of the answer are not in error, and we therefore have three significant digits of accuracy.

The actual source of error in this program is called *roundoff error*, and it is an error due both to the faults of floating-point arithmetic and to the fact that most simple decimal fractions become repeating fractions in the binary number system and therefore cannot be expressed exactly in *any* finite number of bits.

In our sample program, we continually added 0.0001 to itself, 10,000 times, hoping to get the sum 1.00000; let us examine what happens in binary when we add, for example, the decimal 0.0001 to 0.9900 on the 9,900th addition (assuming no previous errors). As described earlier, the decimal 0.0001 is the binary fraction

$$.0000000000000110100011011011100010111010110000\ldots$$

This fraction makes a bad mantissa in a floating-point number because it has 13 initial zeroes, and so it would be normalized into

$$.110100011011011100010111 \times 2^{-13}.$$

Note that only the first 24 bits of the normalized mantissa are kept by our 32-bit computer.

Similarly, the decimal fraction 0.9900 is a binary

$$.1111110101110000101000111101011100001010001111\ldots.$$

Since this is already normalized, it can be stored as the binary floating-point number

$$.111111010111000010100011 \times 2^{0}.$$

Thus the computer's problem is now one of doing the following addition:

$$.110100011011011100010111 \times 2^{-13}$$
$$+.111111010111000010100011 \times 2^{0}$$

Unfortunately, these numbers cannot be added because the two characteristics are not the same. This is analogous to the problem of adding the following two numbers in decimal:

$$0.9900$$
$$+0.0001$$

Before adding the two decimal numbers we must align the decimal points above each other; in the same way, before adding the two floating-point binary numbers, the computer must first *un*-normalize the first number; the addition then becomes

$$.000000000000011010001101 \times 2^{0}$$
$$+.111111010111000010100011 \times 2^{0}$$

$$.111111010111011100110000 \times 2^{0}.$$

Converting this number back to decimal we get 0.99009990 instead of 0.9901, an error of 0.0000001; note that this represents an error in the seventh digit of the answer, so that the answer is accurate to only six digits despite the so-called "seven-digit accuracy" of the computer. Repeat this calculation a few hundred times, and the errors add up.

This brings us to the following definition:

Definition Roundoff *error is the error produced by not using all the bits of a binary number.*

The word *roundoff* is not always accurate in this connection, and comes from the fact that many computers, when cutting off the last digits from a number "round" the number to its closest equivalent. A decimal example is rounding the number 0.358 to 0.36 when dropping the last 8. Some computers do not round, however, but merely cut off the last bits; the resulting error is still properly called roundoff error.[3]

In the above example, roundoff error occurred not only because both binary numbers contained more bits than could be stored in a 24-bit mantissa, but also because of the need to un-normalize a number before addition.

1.6 Absolute and Relative Error

The following definitions apply to errors in a calculation:

Definition *The* absolute *error is defined as*

$$Error_{abs} = (Calculated\ value) - (True\ value)$$

Definition *The* relative *error is defined as*

$$Error_{rel} = \frac{(Calculated\ value) - (True\ value)}{(True\ value)}$$

Definition *The* percentage *error is defined as*

$$Error_{pct} = Error_{rel} \times 100$$

---

[3] The term *truncation* error, sometimes used instead of roundoff error, has a different meaning as we shall see in Chapter 2, and is the wrong word for this error.

EXAMPLE: Suppose the true value for a calculation should be 5.0, but the calculated value is 4.0:

$$\text{Error}_{abs} = 4.0 - 5.0 = -1.0,$$

$$\text{Error}_{rel} = \frac{4.0 - 5.0}{5.0} = \frac{-1.0}{5.0} = -0.2,$$

$$\text{Error}_{pct} = -0.2 \times 100 = -20\%.$$

1.7 Causes of Serious Computer Errors

There are many possible things which can go wrong in a computer program, and we can now discuss some of the more serious. Having had enough of binary numbers, let us switch back to the decimal number system. For the remainder of this chapter, let us consider an imaginary decimal computer with a word length just large enough for a four-digit decimal mantissa, and a one-digit decimal characteristic. We shall assume that this computer might store numbers such as these:

$$.1000 \times 10^1 = 1.0$$
$$.3215 \times 10^{-2} = 0.003215$$
$$.9999 \times 10^9 = 999,900,000$$
$$-.1336 \times 10^{-5} = -0.000001336$$

Using this imaginary computer, we can now examine some possible computer errors:

1. Roundoff in adding or subtracting one large and one small number. We have already seen what happens when we try to add 0.9900 and 0.0001 in a binary computer; this is a general example of what happens when we try to add or subtract two very different numbers.

Suppose we try to add 0.01 to 5000 in our imaginary computer.

$$0.01 = .1000 \times 10^{-1}$$
$$5000 = .5000 \times 10^4.$$

But these normalized numbers cannot be added directly, and so the computer must unnormalize them before doing the addition:

$$
\begin{array}{r}
.000001 \times 10^4 \\
+.5000 \quad\ \times 10^4 \\
\hline
.500001 \times 10^4.
\end{array}
$$

But since the computer can only carry a four-digit mantissa, the last two digits are cut off, and the answer becomes .5000 × 10^4, or 5000. We might just as well not have done the addition at all.

This type of roundoff error is quite common, and can only be avoided by not adding or subtracting two such widely different numbers. See Example 1-1 for a sample of how to do this.

2. Roundoff in subtracting two almost equal numbers. Suppose our imaginary computer is to subtract .1234 from .1235:

$$.1235 \times 10^0$$
$$-.1234 \times 10^0$$

$$\overline{.0001 \times 10^0}.$$

But the mantissa is unnormalized, and so the computer normalizes the answer to .1000 × 10^{-3}.

Both of the two original numbers may have had four significant digits, but the answer only has one significant digit, and we should not rely too much on its accuracy since only a slight error in one of the two original numbers could produce a very large relative error in the answer.

Suppose that the following innocent FORTRAN statement is located in a program:

X = (A − B) ∗ C

There is nothing wrong with it. But suppose that we have the following condition:

$$A = .1235 \times 10^0$$
$$B = .1234 \times 10^0$$
$$C = .1000 \times 10^5 = 10,000.$$

Doing the calculation, we get the result that X = 1, which is quite correct. But suppose that somewhere in an early part of the program the value of A was calculated with a slight roundoff to be .1236 × 10^0. The error so far is very small; the absolute error is only 0.0001, and the relative error is only 0.00081, with a percentage error of only 0.081 percent. But now we use this slightly wrong value of A in the calculation

X = (A − B) ∗ C

and get the answer that X = 2. A tiny error of 0.081 percent has suddenly become a 100 percent error, and the final answer means absolutely nothing.

The problem here is that a computer may make this error completely silently and without warning in the midst of a complex problem, and we may never know. Example 1-2 shows a severe case of this kind of error.

3. Overflow and underflow. Quite often an arithmetic operation with two valid numbers results in an answer which is either too large or too small to be expressed in the computer; this is said to result in *overflow* or *underflow*, respectively.

Still considering the four-digit mantissa and one-digit characteristic of a decimal computer as an example, suppose we multiply

$$.1000 \times 10^6$$
$$\times .2000 \times 10^7$$
$$\overline{\phantom{.0200 \times 10^{13}}}$$
$$.0200 \times 10^{13} = .2000 \times 10^{12}.$$

Although each of the original numbers was well within the reach of a one-digit characteristic, the product is too large, and results in an overflow. Whether we know of such an error depends on the computer system we use— many computers will provide an error message, while others may merely ignore this error and go on, giving the wrong answer. The moral of this entire chapter is "Know thy Computer."

Another case of overflow can happen on the division

$$\frac{1,000,000}{-0.000001} = \frac{.1000 \times 10^7}{-.1000 \times 10^{-5}} = -10^{12} = -.1000 \times 10^{13}.$$

This case of overflow is more often called a *divide check*, as is attempted division by zero, and again the computer will sometimes (but not always) signal such an error. Other cases of overflow appear in the addition or subtraction of very large numbers.

Underflow appears in multiplication or division, and is generally not as serious as overflow; moreover, the computer almost never signals an underflow error. For example,

$$(.1000 \times 10^{-6}) \times (.2000 \times 10^{-5}) = .0200 \times 10^{-11} = .2000 \times 10^{-12}.$$

Since the exponent of -12 is over one digit long, it cannot be expressed in the computer word; hence the number is too small to be stored, and the computer generally makes the answer zero. As an absolute error this is a very small one, and often not serious at all.

But consider the following example. Suppose

$$A = .1000 \times 10^{-6}$$
$$B = .2000 \times 10^{-5}$$
$$C = .3000 \times 10^7,$$

and the computer is attempting to perform

$$X = A * B * C$$

Since the order of precedence for FORTRAN multiplications is from left to right, the multiplication A * B will be done first, and will give an answer of zero. Multiplying by C then still provides an answer of zero. A simple rearrangement of the statement to

$$X = A * C * B$$

will multiply A by C first, and will give the answer $.3000 \times 10^0$ (except for a slight roundoff error). The further multiplication by B will then provide the correct answer of $.6000 \times 10^{-6}$. A similar condition can appear with division, where a slight rearrangement can often avoid an underflow.

4. *Division by a very small number.* As mentioned before, division by a very small number can easily cause an overflow or divide check error. But even a less serious error can cause trouble.

Suppose that a valid division without a divide check is done by the computer and no error is made here. Instead, suppose that a small roundoff error occurred earlier in the program, in calculating the denominator of the division. If the numerator is large and the denominator is small, this can cause a large absolute error in the quotient. If the quotient in turn is subtracted from another number of the same relative size, this can lead to a huge error in the final answer.

As an example, consider the following FORTRAN statement:

$$X = A - B/C$$

where

$$A = .1120 \times 10^9 \quad = 112{,}000{,}000$$
$$B = .1000 \times 10^6 \quad = 100{,}000$$
$$C = .9000 \times 10^{-3} = 0.0009.$$

If the calculation is done on the four-digit decimal computer, the quotient B/C is $.1111 \times 10^9$, and X is $.0009 \times 10^9$ or, after being normalized, X is $.9000 \times 10^6$. Note that there is only one significant digit.

But now suppose that a slight roundoff error in calculating C in some previous step resulted in C being $.9001 \times 10^{-3}$. As an absolute error this is a very small one—only about 10^{-7}; even the relative error is only about 10^{-4} or about 0.01 percent.

But if we now calculate the ratio B/C we get $.1110 \times 10^9$ so that X is now

$.1000 \times 10^7$. That is, the correct X should be 900,000 but is now 1,000,000. As for the errors,

$$Error_{abs} = 1,000,000 - 900,000$$
$$= 100,000$$

$$Error_{rel} = \frac{1,000,000 - 900,000}{900,000} = 0.11$$

$$Error_{pct} = 0.11 \times 100 = 11\%.$$

Thus the relative error has been multiplied by about 1100 times. As before, several such calculations in a program can make the final result absolutely meaningless.

5. *Roundoff in simple multiplication or division.* Whenever we perform a multiplication or division, the result generally has more digits than either original number. For example, when a four-digit number is multiplied by another four-digit number, the result may be either 7 or 8 digits long:

$$3062 \times 5591 = 17119642$$

In our four-decimal-digit computer, however, the answer is rounded to $.1711 \times 10^8$ (or $.1712 \times 10^8$, depending on the exact construction). In either case, we have a roundoff error since we must, of necessity, drop the right half of the number. Even if both of the original numbers are exact, the product has an error.

The same applies to division, as in

$$\frac{.1000 \times 10^1}{.3000 \times 10^1} = .3333 \times 10^0.$$

The correct answer is really a repeating decimal, but only the first four digits can be kept.

6. *Quantizing error.* Since not all numbers can be expressed exactly in a limited number of binary bits, it often happens that the correct answer to a problem cannot be expressed exactly as a binary floating number. This happens more often with large numbers than with small numbers, since large numbers generally require more digits for their accurate expression.

As long as we recognize the fact that the answer may be slightly incorrect, this generally does not bother us. But sometimes the exact answer is important if the computer performs a test with a FORTRAN IF statement on the answer and a match is required.

As an example, a simple procedure for finding the square root of a number

X is to divide it by a guess G and obtain a quotient Q (assuming we don't wish to use the SQRT function):

$$Q = X / G$$

If the original guess G was too large for the square root, then the quotient Q will be too small, and vice versa. In any case, the average of G and Q is probably closer to the actual square root of A than the previous guess, and so we use this as the next guess G.

Obviously this can go on forever, and so there must be some way of deciding when to stop. Clearly, when we finally arrive at the correct square root G,

$$Q = X / G = G,$$

and so we merely test whether Q equals G, and stop if it does.

Unfortunately (see Example 1-3), it may happen that the exact square root cannot be expressed exactly as a binary floating-point number, so that Q and G are never equal, and the program never stops.

The error caused by the fact that a needed number cannot be expressed exactly as a binary floating-point number is called *quantizing* error from the fact that the expressible numbers do not form a continuous set, but instead appear in discrete steps or *quanta*. These steps are larger for large numbers than for small numbers.

7. Output Error. Even if no error is made during the calculation phase of a program, an error can still occur when the results are printed.

As an example, suppose that the answer to a particular calculation is exactly 0.015625; this number happens to be exactly 1/64, and so can be expressed exactly as a finite binary fraction.

As long as this number is printed out with a FORMAT specification such as F10.6 or E14.6, we get the exact answer and have no problem.

But suppose we decide to use an F8.3 code instead. Some computers will round the number and print it as 0.016, while others will print the number as 0.015 without rounding. Prior knowledge of such computer quirks can lead us to anticipate such faults, by a procedure called *half-adjusting*.

To see how this works, let us suppose that the number 5.1193 is to be printed with an F6.2 code in our FORMAT statement. If the computer does not round automatically, the answer will be printed as 5.11. In order to half-adjust, we add a 5 to the first digit to be dropped during printing; in this case we add a 0.005 to the number to get

$$\begin{array}{r} 5.1193 \\ + \ 0.005 \\ \hline 5.1243. \end{array}$$

Printing with an F6.2 code now drops the last two digits, and gives us the desired answer of 5.12, which is better than 5.11.

1.8 Example 1-1. Roundoff Error When Adding a Large Number to a Small Number

The infinite series

$$S = \sum_{n=0}^{\infty} \frac{1}{3^n} = 1 + \tfrac{1}{3} + \tfrac{1}{9} + \tfrac{1}{27} + \tfrac{1}{81} + \tfrac{1}{243} + \cdots$$

has the sum 1.5, as we can easily show from the following. The original series

$$S = 1 + \tfrac{1}{3} + \tfrac{1}{9} + \tfrac{1}{27} + \tfrac{1}{81} + \cdots$$

can be multiplied by 3 on both sides to give

$$3S = 3 + 1 + \tfrac{1}{3} + \tfrac{1}{9} + \tfrac{1}{27} + \tfrac{1}{81} + \cdots$$

$$= 3 + S,$$

since everything after the first term on the right gives the original series. Hence we can solve for S:

$$3S = 3 + S$$
$$2S = 3$$
$$S = 3/2 = 1.5.$$

As an experiment, let us write a short program to sum the first 20 terms of the series, to see whether the sum is really equal to 1.5 as we expect.

We have already found that whenever we add a very large number to a very small one, we automatically make an error. In adding such a series as this, we will always be adding a relatively large number to a relatively small one. For example, after adding the first five terms we already have a sum which is almost equal to 1.5, to which we are adding the number $1/3^5$, which is rather small by comparison. Later on, by the time we get to the $1/3^{19}$ term, the problem is even worse. Hence we do not expect too accurate an answer.

But, as we shall see, there is a trick we can try. Suppose we add the series backwards, *small terms first*. At any particular time, we are adding a particular term in the series to a sum which itself is of the same order of magnitude. By the time we get to the $1/3^1$ term, for example, we already have a sum which equals $\tfrac{1}{6}$. We therefore suspect that summing backward will give us a more accurate answer.

The following program starts by calculating all 20 terms at the start of the series first, and then adds these terms left-to-right, and right-to-left, to compare the results:

```
      DIMENSION TERM (20)
C     CALCULATE ALL THE TERMS OF THE SERIES 1 + 1/3 + 1/9 + 1/27 + ...
      TERM (1) = 1.
      DO 100 I = 2, 20
100   TERM (I) = TERM (I - 1) / 3.
      SUM = 0.
      DO 300 J = 1, 20
300   SUM = SUM + TERM (J)
      WRITE (3, 310) SUM
310   FORMAT (' SUMMING FORWARD WE GET THE ANSWER ',F15.8)
      SUM = 0.
      DO 400 J = 1, 20
      K = 21 - J
400   SUM = SUM + TERM (K)
      WRITE (3, 410) SUM
410   FORMAT(' SUMMING BACKWARD WE GET THE ANSWER',  F15.8)
      CALL EXIT
      END
```

Note that in each case we are adding exactly the same 20 numbers, the only difference being the order of addition. And yet,

```
SUMMING FORWARD WE GET THE ANSWER     1.49999833
SUMMING BACKWARD WE GET THE ANSWER    1.50000000
```

The only way to explain the difference is in terms of roundoff error. Though often there is nothing we can do about roundoff error, this example shows that sometimes a little prior planning can help.

1.9 Example 1-2. Roundoff Error When Subtracting Two Nearly Equal Numbers

Among all the differentiation formulas in any calculus text is

$$\frac{d}{dx} \ln x = \frac{1}{x}.$$

It is interesting to try to check whether this formula is correct by a numerical example, using a computer program.

The derivative is defined as

$$\frac{d}{dx} f(x) = \lim_{\Delta x \to 0} \frac{f(x + \Delta x) - f(x)}{\Delta x}.$$

In the case of ln x, we may write this as

$$\frac{d}{dx} \ln x = \lim_{\Delta x \to 0} \frac{\ln (x + \Delta x) - \ln x}{\Delta x}.$$

The following simple program uses the defining equation to calculate this derivative for the particular case of $x = 3$. Using values of Δx of 0.1, 0.01, etc. down to 10^{-10}, the derivative calculated from the definition (CALCD) is compared with the derivative TRUED calculated from $d \ln x = 1/x = \frac{1}{3}$.

```
      WRITE (3, 10)
10 FORMAT (9X, 'DX', 11X, 'CALC DERIV', 11X, 'ERROR')
      X = 3.
      TRUED = 1. / X
      DO 20 I = 1, 10
      DX = 1. /  (10. ** I)
      CALCD = (ALOG(X+DX) - ALOG(X)) / DX
      ERROR = TRUED - CALCD
20 WRITE (3, 30) DX, CALCD, ERROR
30 FORMAT (E17.8, F15.7, E20.8)
      CALL EXIT
      END
```

Upon running the program we obtain the following results:

DX	CALC DERIV	ERROR
0.10000000E 00	0.3278982	0.54351068E-02
0.10000000E-01	0.3327789	0.55436207E-03
0.10000000E-02	0.3332775	0.55825104E-04
0.10000000E-03	0.3333296	0.36708079E-05
0.10000000E-04	0.3333203	0.12984499E-04
0.10000000E-05	0.3324821	0.85117458E-03
0.10000000E-06	0.3352761	-0.19427933E-02
0.10000000E-07	0.2793967	0.53936561E-01
0.10000000E-08	0.0000000	0.33333333E 00
0.10000000E-09	0.0000000	0.33333333E 00

Since the derivative is defined as the *limit* as $\Delta x \to 0$, we would normally expect the error to decrease for each smaller DX in the table. This indeed happens for the first four lines. But then, somehow, the error again increases until, at the last two lines, the calculated derivative is 0.0 (this program, incidentally, was run in extended precision of about 10 digits.)

The explanation, of course, is in the subtraction

$$\ln (x + \Delta x) - \ln x.$$

When Δx becomes very small, $\ln (x + \Delta x)$ and $\ln x$ are almost equal; when they are subtracted we have a loss of accuracy. When, for example, Δx becomes equal to 10^{-9} (the next-to-last line in the printout), both $\ln (x + \Delta x)$ and $\ln x$ are approximately 1.1, but the difference between them is only about 0.33×10^{-9}. That is, the first ten digits of both are the same, and the eleventh digit is the first where they differ. But if the computer is only capable of storing numbers to about ten digits accuracy, then *as far as the computer is concerned* the numbers are the same, since only the first ten identical digits are stored. The subtraction then gives a result of zero, and division by Δx gives a zero derivative.

We see here an example of an extremely dangerous procedure—the attempt to find the derivative of a function using the defining equation. Differentiation is one of the most (unexpectedly) difficult operations we can try on a computer; in fact, we will not encounter a way out of this problem until the very last pages of this book.

1.10 Example 1-3. Example of Quantizing Error

Quantizing error comes from the fact that the numbers expressible in the binary number system do not form a continuous set, but instead appear in discrete steps or quanta. The following program attempts to find the square root of 1.69; we know that 1.3 is the right answer, but the fractional part 0.3 is a repeating fraction in the binary number system, and so we expect some trouble.

The following program starts with an initial guess of 5, and divides this guess into 1.69 to obtain a quotient of 0.338. Note that the initial guess (5) is too large to be the square root, whereas the quotient (0.338) is too small; we therefore take the average of the two (2.669) as the next guess, and repeat the process. By repeating this process a number of times we hope to get to the point where the guess and the quotient are the same, indicating that we have found the right answer.

The following program, as a precaution, prints the value of the guess at each repetition, so that we can see how the program is doing. When the guess equals the quotient, the program is supposed to stop and print the final answer.

```
C       INITIALIZE
        A = 1.69
        G = 5.0
C       CALCULATE NEW GUESS
   10 Q = A / G
        IF (Q - G) 20, 30, 20
   20 G = (G + Q) / 2.
        WRITE (3, 25) G, Q
   25 FORMAT (2F15.10)
        GO TO 10
   30 WRITE (3, 35) A, G
   35 FORMAT (' THE SQUARE ROOT OF', F10.6, ' IS', F15.7)
        CALL EXIT
        END
```

When we actually run the program, however, we receive the following printout:

```
2.6690001552    0.3380000002
1.6510980157    0.6331959976
1.3373296293    1.0235614804
1.3005211381    1.2637126473
1.3000001935    1.2994792489
1.2999999551    1.2999999551
1.2999999551    1.3000001935
1.2999999551    1.3000001935
1.2999999551    1.3000001935
1.2999999551    1.3000001935
1.2999999551    1.3000001935
1.2999999551    1.3000001935
1.2999999551    1.3000001935
```

The program is stuck in a loop and would continue printing the same line over and over if it were not stopped by the operator. The left column represents the guess G and the right column represents the quotient Q calculated in statement 20 of the program.

The reason for the failure of the program is that, because of quantizing error, the decimal number 1.30000000 cannot be exactly expressed as a binary floating-point number. As a result, the quotient calculated in step 10 of the program is 1.3000001935, the next number above 1.3 that *can* be exactly expressed on this particular computer. When averaged with the previous guess of 1.2999999551, the new guess should be 1.3000000743; this number cannot be exactly expressed, however, and so a slight roundoff is made—right back to 1.2999999551! And so the loop starts again. Since the guess G and the quotient Q are never *exactly* equal, the loop can never stop.

The solution to the problem is to test, not for *exact* equality of Q and G, but for *almost exact* equality. We might, for example, require that

$$|Q - G| \leqslant 0.000001,$$

which means we will accept the answer if Q and G are within 0.000001 of each other.

This is a dangerous procedure, however, since it will work only for small square roots. If we should try to find the square root of 1,334,526 (for example), Q and G would both be over 1000 in magnitude, and quantizing error might make it impossible for them to be that close to each other.

An alternative solution is to require that the difference between the quotient Q and the guess G must be less than some fraction of G; this is equivalent to saying we will accept a small relative error even though the absolute error may be large.

Making use of the ABS function in FORTRAN to take the absolute value of $|Q - G|$, we write the following IF statement:

```
IF (ABS(Q - G) - 0.000001 * G) 30, 20, 20
```

When we run the program again, we obtain the answer

```
2.6690001552    0.3380000002
1.6510980157    0.6331959976
1.3373296293    1.0235614804
1.3005211381    1.2637126473
1.3000001935    1.2994792489
THE SQUARE ROOT OF  1.690000 IS        1.3000001
```

The loop has repeated itself five times, and finally the IF statement has found a match between Q and G, and the program prints the final answer and halts.

Note that there is a price to pay in exchange, though. We still cannot express the final answer exactly, and must accept the answer 1.3000001 despite its slight error.

1.11 Propagation of Errors

Once we know how errors originate within a computer program, we are tempted to try to analyze each step within a program to determine the error made in that step, and in this way determine the total error in the final answer. If this were possible, we could apply an appropriate correction factor to the computed answer, and thus be assured of an exact result. Unfortunately, this is not a practical approach because it would involve as much work as solving the entire problem by hand.

Nevertheless, we can try to analyze individual computer operations, as we have already done, to see when errors arise. We will now provide a small introduction to the mathematical analysis of such errors for the four basic arithmetic operations of addition, subtraction, multiplication, and division. In each case we shall assume that the operation is to be carried out on two numbers a and b.

1. Addition. In adding the two numbers a and b, we expect the correct sum to be $a + b$; in general, however, we obtain a sum which has an error caused by roundoff. To account for this error, we may consider it as being caused by the computer performing a faulty operation $\dot{+}$ (where the dot indicates that an error is made) instead of the correct operation $+$. We can then interpret the roundoff error as

$$\text{Error} = (a \dot{+} b) - (a + b).$$

The magnitude of this error depends on the relative magnitudes and the signs of a and b, and we need to actually consider the bit structure of a and b as they are stored in the computer.

There is a second error possible, however, if a and b themselves are inaccurate to begin with. Consider, for example, that instead of the true value of a the computer has the value $\dot{a}$ which is in error by an amount ε_a

$$\dot{a} = a + \varepsilon_a$$

and similarly

$$\dot{b} = b + \varepsilon_b.$$

Then, even if no further roundoff error occurred during the addition, the sum would still have an error (called a *propagation error*):

$$\begin{aligned} \text{Error} &= (\dot{a} + \dot{b}) - (a + b) \\ &= (a + \varepsilon_a + b + \varepsilon_b) - (a + b) \\ &= \varepsilon_a + \varepsilon_b. \end{aligned}$$

In the complete case, where we have both the inaccurate values $\dot{a}$ and $\dot{b}$ as well as an erroneous operation $\dot{+}$, the error becomes

$$\text{Error} = (\dot{a} \dot{+} \dot{b}) - (a + b),$$

where the operation $\dot{+}$, since it depends on the specific values of the numbers being operated on, can be considered as nonlinear. That is, we are *not* allowed to interpret the term $(\dot{a} \dot{+} \dot{b})$ as

$$\begin{aligned} (\dot{a} \dot{+} \dot{b}) &= (a + \varepsilon_a) \dot{+} (b + \varepsilon_b) \\ &\overset{?}{=} (a \dot{+} b) + (\varepsilon_a \dot{+} \varepsilon_b). \end{aligned}$$

Hence it is easy to mathematically consider the effect of inaccurate values of $\dot{a}$ and $\dot{b}$ on the result, but difficult to consider the effect of $\dot{+}$ as well. Thus propagation error is interpreted as the error caused by inaccurate starting values in a computation *propagating* through the computation and causing an inherent error in the result.

 2. *Subtraction.* The propagation error caused by inaccurate starting values of $\dot{a}$ and $\dot{b}$ can be obtained from the case of addition by merely changing the sign; the effect of the subtraction operation $\dot{-}$ on accuracy is also similar to the effect of the addition operator $\dot{+}$ except for a change in sign.

 3. *Multiplication.* If we multiply the numbers $\dot{a}$ and $\dot{b}$ we obtain (neglecting error caused by the operator $\dot{*}$ itself)

$$\begin{aligned} \dot{a} * \dot{b} &= (a + \varepsilon_a) * (b + \varepsilon_b) \\ &= (a * b) + (a * \varepsilon_b) + (b * \varepsilon_a) + (\varepsilon_a * \varepsilon_b). \end{aligned}$$

If we assume that ε_a and ε_b are sufficiently small, we may consider their product to be small compared with the other terms and thus drop the last term. We then obtain the error in the overall result as

$$(\dot{a} * \dot{b}) - (a * b) \approx (a * \varepsilon_b) + (b * \varepsilon_a).$$

It is now possible to find the relative error in the result by dividing both sides by $a * b$:

$$\frac{(\dot{a} * \dot{b}) - (a * b)}{(a * b)} \approx \frac{\varepsilon_b}{b} + \frac{\varepsilon_a}{a}.$$

Hence the relative propagation error caused during multiplication (ignoring the additional error caused by the faulty operation of $\dot{*}$) is simply the sum of the relative errors inherent in $\dot{a}$ and $\dot{b}$.

As in addition and subtraction, the additional error caused by the operation $\dot{*}$ requires detailed analysis of the exact way the particular numbers $\dot{a}$ and $\dot{b}$ are stored, and the analysis of this error is really not practical.

4. Division. Neglecting the fact that division of two numbers causes an additional roundoff error which we may interpret as being caused by the faulty operation $\dot{/}$, we may consider the division of $\dot{a}$ by $\dot{b}$ as follows:

$$\dot{a}/\dot{b} = (a + \varepsilon_a)/(b + \varepsilon_b)$$

$$= (a + \varepsilon_a)\frac{1}{(b + \varepsilon_b)}.$$

The fraction on the right can be expanded by long division into a series:

$$\frac{1}{b} - \frac{\varepsilon_b}{b^2} + \frac{\varepsilon_b^2}{b^3} - \frac{\varepsilon_b^3}{b^4} + \cdots$$

$$b + \varepsilon_b \overline{)\,1}$$

$$1 + \frac{\varepsilon_b}{b}$$

$$-\frac{\varepsilon_b}{b}$$

$$-\frac{\varepsilon_b}{b} - \frac{\varepsilon_b^2}{b^2}$$

$$+\frac{\varepsilon_b^2}{b^2}$$

$$+\frac{\varepsilon_b^2}{b^2} + \frac{\varepsilon_b^3}{b^3}$$

$$-\frac{\varepsilon_b^3}{b^3}$$

$$\vdots$$

We therefore have:

$$\dot{a}/\dot{b} = (a + \varepsilon_a)\left[\frac{1}{b} - \frac{\varepsilon_b}{b^2} + \frac{\varepsilon_b^2}{b^3} - \frac{\varepsilon_b^3}{b^4} + \cdots\right].$$

We now multiply and neglect the higher terms in ε_b:

$$\dot{a}/\dot{b} = \frac{a}{b} + \frac{\varepsilon_a}{b} - \frac{a\varepsilon_b}{b^2} - \frac{\varepsilon_a\varepsilon_b}{b^2} + \text{neglected terms}.$$

By comparison with the other terms, the fourth term on the right can also be deleted, and we have just

$$\dot{a}/\dot{b} \approx a/b + \frac{\varepsilon_a}{b} - \frac{a\varepsilon_b}{b^2}.$$

Thus the error becomes

$$\dot{a}/\dot{b} - a/b \approx \frac{\varepsilon_a}{b} - \frac{a\varepsilon_b}{b^2}.$$

Dividing by a/b to obtain the relative error,

$$\frac{\dot{a}/\dot{b} - a/b}{a/b} \approx \frac{\dfrac{\varepsilon_a}{b} - \dfrac{a\varepsilon_b}{b^2}}{a/b} = \frac{\varepsilon_a}{a} - \frac{\varepsilon_b}{b}.$$

Thus the relative error in the quotient is approximately equal to the difference between the relative errors of the original numbers $\dot{a}$ and $\dot{b}$. This is only expected, since if both $\dot{a}$ and $\dot{b}$ are too high or too low by the same relative amount, then their quotient will be the correct value.

Finally, we may look at the propagated error (assuming ideal operations $+$, $-$, $*$ and $/$) in calculating the value of some function $f(x)$ at the point $x = a$. We hence desire the value $f(a)$; in general, however, we may have available only an erroneous value $\dot{a}$ instead of the real value of a; what is the resulting error in $f(\dot{a}) - f(a)$?

Until we cover Taylor series in Chapter 2, we shall attack the problem graphically (see problem 17 at the end of Chapter 2 for an analytic proof). Figure 1-1 shows the function $f(x)$ plotted in the vicinity of $x = a$. Between $x = a$, the true value, and $x = \dot{a}$, the value available, there is an error ε_a. Corresponding to this error exists another error ε_f between the true $f(a)$ and the calculated $f(\dot{a})$. What is the relation between ε_a and ε_f?

If ε_a is small, we can approximate the curve $f(x)$ by its tangent at $x = a$.

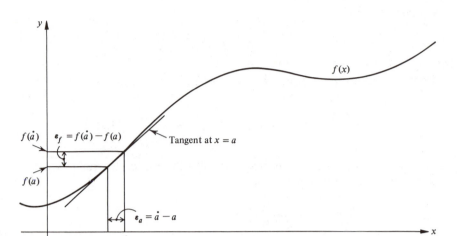

Figure 1-1.   Derivation of error in calculating $f(\dot{a})$ instead of $f(a)$.

We then know that the slope of this tangent is $f'(a)$, and also that its slope is $\varepsilon_f/\varepsilon_a$; hence we can solve for ε_f :

$$\varepsilon_f/\varepsilon_a \approx f'(a)$$

$$\varepsilon_f \approx \varepsilon_a f'(a) \approx \varepsilon_a f'(\dot{a}).$$

Hence the error in evaluating a function based on inaccurate knowledge of its argument is proportional to the first derivative of the function at the point where it is being evaluated.

1.12 Problems

1. When comparing the decimal and binary number systems, we say that the decimal number system uses the *base* 10, while the binary number system uses the *base* 2. How do we interpret the meaning of the word *base*?

2. Convert the following binary numbers to decimal:

   a. 11011
   b. 11111
   c. 11.001
   d. 0.101

3. In the binary number system, what is the next larger integer after 11; after 111; after 11111?

4. Based on the results of problem 3, you are given the following question: The largest number that can be stored in a particular computer is the binary integer 11111111111111111111, which contains twenty ones. Find an easy way to convert this into decimal without rewriting it as

$$(1 \times 2^{19}) + (1 \times 2^{18}) + \cdots + (1 \times 2^1) + (1 \times 2^0).$$

5. Moving the radix point in a binary number n places left (or right) is the same as dividing (or multiplying, respectively) the number by 2^n. Justify this statement.

6. Based on the above problem, rewrite the binary fraction 0.11011 as a binary integer times an appropriate power of 2. Use this operation to prove the statement, "A rational fraction can be exactly expressed with a finite number of binary digits only if it can be expressed as the quotient of two integers p/q, where q is a power of 2, such as $q = 2^n$ for some integer n."

7. Prove to yourself that every number of the form 2^{-n} where n is a positive integer can be expressed as a decimal number with a finite number of digits.

8. Show that any binary fraction which is finite is also a finite fraction if converted to a decimal fraction.

9. Show that any decimal fraction which is repeating cannot be finite when we convert it to binary. What does this mean in connection with irrational numbers?

10. A 32-bit computer has the following floating-point number:

00000110010100100111011011110110.

If the first bit is the over-all sign, the second bit is the sign of the characteristic, the next six bits are the characteristic, and the last 24 bits are the mantissa, answer the following questions:

a. Is the number normalized? If not, normalize it.
b. Is it positive or negative?
c. Is it greater than 1 or less than 1 in magnitude?

11. Repeat question 10 for the number

00000000101010101010101010101010

12. In the 32-bit computer in problem 10, what is the approximate largest number that can be stored? What is the largest positive integer that can can be stored *exactly*?

13. A student used the following steps inside a program:

```
X = 0.0
DO 10 I = 1, 10
   X = X + 0.1
10 WRITE (3, 20) X
20 FORMAT (F6.1).
```

Upon running the program he got the following table:

$$
\begin{array}{c}
0.1 \\
0.2 \\
0.2 \\
0.3 \\
0.4 \\
0.5 \\
0.6 \\
0.7 \\
0.8 \\
0.9
\end{array}
$$

What happened, and why?

14. Write a program to sum the series

$$4 - \tfrac{4}{3} + \tfrac{4}{5} - \tfrac{4}{7} + \tfrac{4}{9} - \cdots,$$

starting from the left until a term less than 0.00001 is encountered. Then print the sum, and recalculate the terms backward from the right and sum again; compare the results.

15. Two numbers 3.141 and 57.295 each have a maximum error of 0.001. What is the maximum *relative* error in each number? What is the maximum relative error in their sum? difference? product? quotient?

16. In evaluating a function $f(x) = e^{-x} \sin x$ at $x = 1$, what is the error in $f(x)$ if the x actually used in the computation is in error by 0.01?

17. In Example 1-1 we summed the first 20 terms of the series

$$S = \sum_{n=0}^{\infty} \frac{1}{3^n} = 1 + \frac{1}{3} + \frac{1}{9} + \frac{1}{27} + \frac{1}{81} + \frac{1}{243} + \frac{1}{729} + \cdots.$$

Now define the partial sum S_m which includes only the first $m + 1$ terms:

$$S_m = \sum_{n=0}^{m} \frac{1}{3^n} = 1 + \frac{1}{3} + \frac{1}{9} + \frac{1}{27} + \cdots + \frac{1}{3^m}.$$

Then in the limit we have $\lim_{m \to \infty} S_m = 1.5$, as before.

a. If we define $S_0 = 1$, show that

$$S_{m+1} = 1 + \frac{S_m}{3} \qquad \text{for } m = 0, 1, 2, \ldots .$$

b. Use this recurrence equation to compute S_1 through S_{20} and compare with the values obtained in Example 1-1.

c. Repeat part *b* with other choices of S_0, such as $S_0 = 0, 0.5, 1.5$, and 2.

d. Suppose that, in starting with $S_0 = 1$, the computer were to make a 25 percent roundoff error in calculating S_5; how would this affect the value of S_{20}?

2

Power Series Calculation of Functions

The FORTRAN system on a computer generally supplies such elementary functions as the SIN, COS, EXP, and ALOG (natural logarithm) functions, and thus saves us the effort of calculating our own. But since we may often not wish to use the function provided, or may not have the right function available, this chapter will show how these functions are calculated. This is a subject in its own right, since the Taylor series we shall derive in this chapter is also needed for a number of other derivations and proofs.

2.1 Taylor and Maclaurin Series

Suppose we need to know the sine of a particular angle. Our natural impulse is to reach for a book of trigonometric tables and look it up, but such a procedure is very wasteful for a computer because it requires use of valuable computer memory to store large tables.

The above *table-lookup* method, as it is called, essentially amounts to a knowledge of the function's value at a large number of selected points over a range of interest. Another approach is to know everything there is to be known about the function *at one point*, and use this knowledge to calculate the value of the function at *any other needed point*. This is the power series approach.

30

Figure 2-1 shows the graph of some function $y = f(x)$. We are going to assume that this function is continuous and that all its derivatives exist within the range of interest to us. For the elementary functions such as the sine or cosine, this is no problem; for a function such as the natural logarithm we must stay away from the point $x = 0$ since the function itself, as well as all its derivatives, is infinite there. (We shall see how to treat this problem in a moment.)

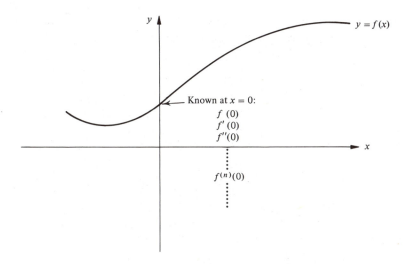

Figure 2-1.   The function $y = f(x)$.

So we consider the function $y = f(x)$ to be continuous and differentiable over some range which, for the present, will include the point $x = 0$ (thus letting out the logarithm for now). Let us assume that we know everything there is to be known about $f(x)$ at the point $x = 0$. What do we mean by the phrase *all there is to be known*? We mean that we know the value of the function at $x = 0$ [i.e., we know the value $f(0)$], we know the slope of the curve through that point [i.e., we know the value of $f'(0)$, the first derivative], we know the curvature of the curve around that point [i.e., we know $f''(0)$, the second derivative], and in fact we know the values of all the derivatives at the point $x = 0$.

Hence we know the following:

$$f\ (0)$$
$$f'\ (0)$$
$$f''\ (0)$$
$$\vdots$$
$$f^{(n)}(0)$$
$$\vdots \quad ,$$

where the symbol $f^{(n)}(0)$ stands for the nth derivative evaluated at the point $x = 0$.

Now suppose that we need to know the value of the function $f(x)$ evaluated at a point x_1, shown in Figure 2-2, which is very close to the origin at $x = 0$. In this case we can make the approximation that

$$f(x_1) \approx f(0)$$

and make only a small error. But obviously this approximation will not work if we need $f(x_2)$, since x_2 is too far away from $x = 0$.

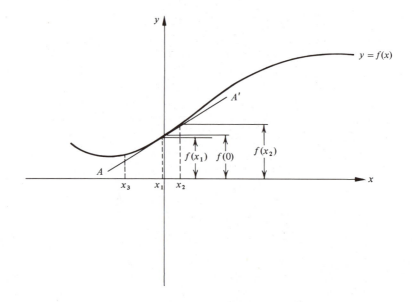

Figure 2-2.   Approximating $f(x_1)$ and $f(x_2)$.

The next best approximation to $f(x_2)$ can be obtained by drawing the tangent line AA' to the curve as shown in Figure 2-2. This line is tangent to the curve at $x = 0$, and has the slope $f'(0)$, equal to the derivative of the function at this point. If x_2 is close to the origin, the tangent line is a good approximation to the curve, and we can read the value of $f(x_2)$ from the tangent line instead of from the curve without too much of an error.

Using the standard form $y = mx + b$ for a straight line, we write

$$y = mx + b$$
$$f(x_2) \approx f'(0)\, x_2 + f(0)$$
$$\approx f(0) + f'(0)\, x_2.$$

In general, for any other point close to 0, we could find $f(x)$ from

$$f(x) \approx f(0) + f'(0)\, x$$

which is a first-order polynomial in x chosen so that its value as well as its first derivative match those of the curve $f(x)$ itself.

While this approximation will work for x_2, it will obviously not work for $f(x_3)$, since x_3 is too far away from $x = 0$, and the tangent line is no longer a good approximation to the curve. By extension of the previous reasoning, we would now like to draw a higher-order curve, a parabola, which would be the next best approximation. As before, its ordinate value as well as its derivatives should match those of the function $f(x)$ at $x = 0$. The resulting equation used as an approximation would be a second-degree polynomial in x,

$$f(x) \approx a_0 + a_1 x + a_2 x^2,$$

where we have already found that

$$a_0 = f(0)$$
$$a_1 = f'(0),$$

and all that remains is to find the constant a_2. If we take the approximating polynomial and differentiate it twice, we may set it equal to $f''(0)$, the second derivative of the original function $f(x)$ evaluated at $x = 0$:

$$f''(0) = 2a_2$$

or

$$a_2 = \frac{f''(0)}{2},$$

so that the second-degree approximating polynomial becomes

$$f(x) \approx f(0) + f'(0)\, x + f''(0)\, \frac{x^2}{2}.$$

We find that the resulting approximation is valid for a slightly greater distance from $x = 0$, but eventually we must improve it even more. In general, we find that if we wish to go infinitely far away from $x = 0$, or if we demand infinite accuracy—the exact value for $f(x)$ at any x—then we must go to a polynomial of infinite degree (which is now an exact equality):

$$f(x) = a_0 + a_1 x + a_2 x^2 + a_3 x^3 + \cdots + a_n x^n + \cdots, \qquad \text{(2-1)}$$

which is essentially an infinite power series in x. We now wish to find the exact value of the coefficient a_n for any term.

If both sides of Eq. (2-1) are to be exactly equal, then all their derivatives must be equal as well. To find the coefficient of the term in x^n, differentiate n times:

$$
\begin{aligned}
f(x) &= a_0 + a_1 x^1 + a_2 x^2 + \cdots + a_n x^n + \cdots \\
f'(x) &= a_1 + 2a_2 x + \cdots + na_n x^{n-1} + \cdots \\
f''(x) &= 2a_2 + \cdots + n(n-1)a_n x^{n-2} + \cdots \\
&\;\;\vdots \\
f^{(n)}(x) &= n!\,a_n \phantom{x^{n-2}} + \cdots,
\end{aligned}
$$

where the higher terms will all be powers of x.[1] We now set x equal to 0, and obtain

$$ f^{(n)}(0) = n!\,a_n, $$

where all the remaining terms drop out. This can be solved for

$$ a_n = \frac{f^{(n)}(0)}{n!}, $$

which gives us an infinitely long power series for $f(x)$:

$$ f(x) = f(0) + f'(0)x + f''(0)\frac{x^2}{2!} + f'''(0)\frac{x^3}{3!} + f^{IV}(0)\frac{x^4}{4!} $$

$$ + \cdots + f^{(n)}(0)\frac{x^n}{n!} + \cdots. \tag{2-2} $$

This series is called a *Maclaurin series*.

The Maclaurin series for $f(x)$ is obtained by evaluating the function $f(x)$ and all its derivatives at $x = 0$. As pointed out earlier, this is often inconvenient or even impossible, as for the natural logarithm which is infinite at $x = 0$. For this reason we have a generalization of the Maclaurin series which, in turn, is called a *Taylor series*.

To develop the Taylor series, let us consider a function $f(x)$, which satisfies our condition of continuity and possesses all the derivatives we need within some region of interest. Let us suppose that we know the value of this function as well as all its derivatives at some point $x = x_0$, where x_0 could be zero (in which case we have a Maclaurin series), or could be some other point.

---

[1] The notation $n!$ is called *factorial n*, and denotes the product $n(n-1)(n-2)\ldots(3)(2)(1)$. Thus, for example, $3! = (3)(2)(1) = 6$.

We now form an infinite power series for $f(x)$ as before, except that we use powers of $(x - x_0)$ instead of powers of x:

$$f(x) = a_0 + a_1(x - x_0) + a_2(x - x_0)^2 + \cdots + a_n(x - x_0)^n$$
$$+ a_{n+1}(x - x_0)^{n+1} + \cdots .$$

As before, we differentiate n times, and have

$$f^{(n)}(x) = n!a_n + (n + 1)!a_{n+1}(x - x_0) + \cdots .$$

Setting x equal to x_0, the terms in $(x - x_0)$ drop out, and we have

$$f^{(n)}(x_0) = n!a_n$$
$$a_n = \frac{f^{(n)}(x_0)}{n!}$$

for any n. The Taylor series is then

$$f(x) = f(x_0) + f'(x_0)(x - x_0) + f''(x_0)\frac{(x - x_0)^2}{2!} + f'''(x_0)\frac{(x - x_0)^3}{3!}$$
$$+ \cdots + f^{(n)}(x_0)\frac{(x - x_0)^n}{n!} + \cdots . \tag{2-3}$$

Since the Maclaurin series is only a special case of the Taylor series, we often use the name Taylor series for both types, regardless of whether $x_0 = 0$ (Maclaurin) or not (Taylor).

2.2 Examples of Taylor and Maclaurin Series

1. *Series for the sine of an angle x.* If $y = f(x) = \sin x$, we can express the sine in terms of a Taylor series by evaluating the function and all derivatives at some point x_0. If we evaluate at $x_0 = 0$ (this is called expanding in a series about $x_0 = 0$) we get a particularly easy series:

$$f(0) = \sin 0 = 0,$$
$$f'(0) = \cos 0 = 1,$$
$$f''(0) = -\sin 0 = 0,$$
$$f'''(0) = -\cos 0 = -1,$$
$$f^{\text{IV}}(0) = \sin 0 = 0,$$

and the derivatives start to repeat, so that the series becomes a Maclaurin series

$$\sin x = f(0) + f'(0)x + f''(0)\frac{x^2}{2!} + f'''(0)\frac{x^3}{3!} + f^{IV}(0)\frac{x^4}{4!} + \cdots$$

$$= \quad 0 \quad + \quad x \quad + \quad 0 \quad - \quad \frac{x^3}{3!} \quad + \quad 0 \quad + \cdots$$

$$= x - \frac{x^3}{3!} + \frac{x^5}{5!} - \frac{x^7}{7!} + \frac{x^9}{9!} - \cdots.$$

2. Series for the cosine of an angle x. The cosine can also be expanded about $x_0 = 0$ to give the Maclaurin series

$$\cos x = 1 - \frac{x^2}{2!} + \frac{x^4}{4!} - \frac{x^6}{6!} + \frac{x^8}{8!} - \cdots.$$

3. Series for e^x. Since $d/dx\ e^x = e^x$, the function and all its derivatives at $x_0 = 0$ are equal to $e^0 = 1$, so that the Maclaurin series becomes

$$e^x = 1 + x + \frac{x^2}{2!} + \frac{x^3}{3!} + \frac{x^4}{4!} + \frac{x^5}{5!} + \cdots.$$

4. Series for ln x. The sine, cosine, and e^x series were all Maclaurin series because they happened to be particularly easy to expand around $x_0 = 0$. This is not possible for the ln x series, and so we must expand this series for some other x_0. Any positive value would work, except that $x_0 = 1$ happens to be particularly convenient. We proceed as follows:

$$f(x) = \ln x \qquad\qquad f(1) = \ln 1 = 0$$
$$f'(x) = 1/x = x^{-1} \qquad\qquad f'(1) = 1/1 = 1$$
$$f''(x) = -x^{-2} = -1/x^2 \qquad\qquad f''(1) = -1/1 = -1$$
$$f'''(x) = 2x^{-3} = 2/x^3 \qquad\qquad f'''(1) = 2/1 = 2$$
$$f^{IV}(x) = -6x^{-4} = -6/x^4 \qquad\qquad f^{IV}(1) = -6/1 = -6$$
$$f^{V}(x) = 24x^{-5} = 24/x^5 \qquad\qquad f^{V}(1) = 24/1 = 24$$
$$\vdots \qquad\qquad\qquad \vdots$$

The resulting Taylor series becomes

$$\ln x = 0 + 1(x - 1) - 1\frac{(x - 1)^2}{2!} + 2\frac{(x - 1)^3}{3!} - 6\frac{(x - 1)^4}{4!}$$

$$+ 24\frac{(x - 1)^5}{5!} - \cdots.$$

We notice that the coefficients always cancel out part of the factorial in the denominator, so that the series can be simplified to

$$\ln x = (x - 1) - \frac{(x - 1)^2}{2} + \frac{(x - 1)^3}{3} - \frac{(x - 1)^4}{4} - \frac{(x - 1)^5}{5} + \cdots.$$

We now ask an interesting question—just what values of x is this series good for? Obviously, if $x = 0$ the function $\ln x$ becomes $-\infty$; let us insert $x = 0$ into the series to see what happens:

$$\ln 0 \stackrel{?}{=} -1 - \tfrac{1}{2} - \tfrac{1}{3} - \tfrac{1}{4} - \tfrac{1}{5} - \cdots.$$

From our knowledge of series we know that this series does not converge to any finite sum; we might say that its "sum" is $-\infty$, as expected.

What about $\ln 2$? We know that the logarithm exists, but it is interesting to let $x = 2$ in the Taylor series:

$$\ln 2 \stackrel{?}{=} 1 - \tfrac{1}{2} + \tfrac{1}{3} - \tfrac{1}{4} + \tfrac{1}{5} - \cdots.$$

Does this series have a finite sum? A theorem by Leibnitz is of some help here:

Theorem *If an infinite series*

$$c_1 - c_2 + c_3 - c_4 + \cdots$$

satisfies the conditions that
   1. It is strictly alternating,
   2. Each term is smaller in magnitude than the term before it,
   3. The terms approach zero as a limit,
then the series has a finite sum, and moreover, if we stop adding the terms after the nth term, the error thus produced is between zero and the first non-zero term not taken.

The series for $\ln 2$ satisfies these three conditions and therefore has a finite sum; moreover, we can approximate that sum with the first five terms which add up to approximately 0.783, with an error between 0 and $-\tfrac{1}{6}$, the first term not taken. Thus $\ln 2$ must have a value between about 0.617 and 0.783; the correct value is 0.693 approximately.

This brings up the point that an infinite series is useless for purposes of actual computation, and we must generally cut (or *truncate*) the series at some finite point. In the next section we shall discuss the error made by not including all the terms of the series; now we only wish to examine the implications on the $\ln x$ series. Quite a complicated situation arises if we try to find ln x for some x greater than 2. As a concrete example, for ln 3 we have

$$\ln 3 \overset{?}{=} (3 - 1) - \frac{(3 - 1)^2}{2} + \frac{(3 - 1)^3}{3} - \frac{(3 - 1)^4}{4} + \frac{(3 - 1)^5}{5} - \cdots$$

$$\overset{?}{=} 2 \quad - \frac{2^2}{2} + \frac{2^3}{3} - \frac{2^4}{4} + \frac{2^5}{5} - \cdots.$$

Each term is larger than the term before, and the terms keep increasing in magnitude indefinitely, as we can see from a simple analysis: The nth term in the above series is $2^n/n$ in magnitude where n is any positive integer; the next term in the series is then $2^{n+1}/(n + 1)$; the ratio between the $(n + 1)$th term and the nth term is

$$\frac{|c_{n+1}|}{|c_n|} = \frac{\dfrac{2^{n+1}}{n + 1}}{\dfrac{2^n}{n}}$$

$$= \frac{2^{n+1}}{n + 1} \frac{n}{2^n} = 2\frac{n}{n + 1}.$$

As n becomes large, each term is therefore approximately twice as large (in magnitude) as the term before it. Obviously this series is useless for practical computation since it simply does not converge to any finite sum.

Hence we must be very careful when using Taylor series that the original assumptions about the function $f(x)$ which we are approximating are satisfied—$f(x)$ must be a continuous function and have all its derivatives in the region of interest. The difficulty with ln x is that it does not fulfill these conditions at $x = 0$; we say that $x = 0$ is a *singular point* of the function ln x.

In general, when a function $f(x)$ is expanded about some known point at $x = x_0$, the Taylor series expansion for the function is only valid in a region known as the *region of convergence*; outside this region the Taylor series will not converge to a finite sum. The region of convergence, on the other hand, extends in all directions a distance equal to the distance to the closest singular point.

Here is the crux of our problem with the ln x series. When expanded about $x_0 = 1$, the distance to the nearest singular point at $x = 0$ is 1; hence the region of convergence lies only from $x = 0$ to $x = 2$. The Taylor series happens to converge for $x = 2$, but will diverge for any x greater than 2 or for any negative x. In the case of the sine, cosine, and e^x series there were no singular points (except possibly at infinity) and so the region of convergence extended for all x satisfying $-\infty < x < +\infty$.

2.3 Taylor and Maclaurin Series with a Remainder

Each of the series we have considered so far is an infinite series, having an infinite number of terms. In evaluating a series which converges, if we wish an exact answer we must calculate each of the infinite number of terms and add it into the series. Since this would take an infinite time, we can only approximate by taking the first few terms, and neglecting the rest. Fortunately, because of the factorials in the denominator of each term, the later terms in converging series become smaller and smaller so that, even though there is an infinite number of them, their sum is small enough that by neglecting them we make only a small error. This error is called *truncation error* because it occurs as a result of truncating (cutting) the series, and is completely different from roundoff error (which is also present if this calculation is done on a computer).

Suppose we take only the first n terms of a Taylor series, and discard the rest:

$$\underbrace{f(x) = f(x_0) + f'(x_0)(x - x_0) + \cdots + f^{(n-1)}(x_0)\frac{(x - x_0)^{n-1}}{(n - 1)!}}_{\text{Only first } n \text{ terms taken}} + \begin{array}{l} \text{Remaining} \\ \text{Terms} \\ \text{Truncated} \end{array}$$

The terms not taken represent an error in our calculation; let us represent the sum of all the terms not taken by the symbol R_n, meaning the "remainder after n terms".

We may then write

$$\text{actual } f(x) = \text{first } n \text{ terms} + R_n.$$

Our next problem is to find an expression for R_n since its knowledge will be useful in determining the error we have made by truncating everything after the first n terms.

It is not easy to intuitively justify the next step except to say that it gives us a convenient way of deriving R_n. Suppose that we integrate $f'(t)$, where t is a dummy variable substituted instead of x into the expression for the derivative of $f(x)$, between the limits of x_0 and x:

$$\int_{x_0}^{x} f'(t)\, dt = \left[f(t) \right]_{x_0}^{x} = f(x) - f(x_0).$$

Rearranging, we have

$$f(x) = f(x_0) + \int_{x_0}^{x} f'(t)\, dt. \tag{2-4}$$

Comparing with the Taylor series expansion for $f(x)$ about $x = x_0$,

$$f(x) = f(x_0) + f'(x_0)(x - x_0) + f''(x_0)\frac{(x - x_0)^2}{2!} + \cdots. \quad (2\text{-}3)$$

we see that the integral in (2-4) is just the sum of the remaining terms in the series after the first. Hence

$$R_1 = \int_{x_0}^{x} f'(t)\, dt.$$

We now take this integral and integrate by parts, letting

$$u = f'(t) \qquad\qquad dv = dt$$
$$du = f''(t)\, dt \qquad\quad v = t - x.$$

(Since the integration is done with respect to t, we can treat x as a constant of integration and insert it into the expression for v to anticipate a future step.) Integrating, we now have:

$$\int_{x_0}^{x} u\, dv = \left[uv \right]_{x_0}^{x} - \int_{x_0}^{x} v\, du$$

$$\int_{x_0}^{x} f'(t)\, dt = \left[f'(t)(t - x) \right]_{x_0}^{x} - \int_{x_0}^{x} (t - x)f''(t)\, dt$$

$$= f'(x)(x - x) - f'(x_0)(x_0 - x) - \int_{x_0}^{x} (t - x)f''(t)\, dt$$

$$= f'(x_0)(x - x_0) + \int_{x_0}^{x} (x - t)f''(t)\, dt. \quad (2\text{-}5)$$

Inserting back into (2-4), we obtain

$$f(x) = f(x_0) + f'(x_0)(x - x_0) + \int_{x_0}^{x} (x - t)f''(t)\, dt, \quad (2\text{-}6)$$

where the integral on the right is now obviously equal to R_2:

$$R_2 = \int_{x_0}^{x} (x - t)f''(t)\, dt.$$

We once again integrate by parts, letting

$$u = f''(t) \qquad dv = (x - t)\, dt$$

$$du = f'''(t)\, dt \qquad v = -\frac{(x - t)^2}{2}.$$

When we substitute the resulting integral back into (2-6), we obtain

$$f(x) = f(x_0) + f'(x_0)(x - x_0) + f''(x_0)\frac{(x - x_0)^2}{2}$$

$$+ \int_{x_0}^{x} \frac{(x - t)^2}{2} f'''(t)\, dt,$$

where the last integral is R_3. Continuing in this way, we could integrate as many times as we wish, to obtain any R_n:

$$R_n = \int_{x_0}^{x} \frac{(x - t)^{n - 1}}{(n - 1)!} f^{(n)}(t)\, dt. \qquad (2\text{-}7)$$

This form for the remainder R_n is somewhat inconvenient, and so we convert to the more common *Lagrangian* form as follows. During the integration, the dummy variable t goes from $t = x_0$ to $t = x$; at each different t, $f^{(n)}(t)$ has a different value. Let M be the maximum of all these values, and let m be the minimum. We could then form an inequality[2]

$$\int_{x_0}^{x} m\frac{(x - t)^{n - 1}}{(n - 1)!}\, dt \leqslant R_n \leqslant \int_{x_0}^{x} M\frac{(x - t)^{n - 1}}{(n - 1)!}\, dt.$$

Since m and M are constants, they may be brought outside the integral, and the integration is then quite easy:

$$m\left[\frac{-(x - t)^n}{n!}\right]_{x_0}^{x} \leqslant R_n \leqslant M\left[\frac{-(x - t)^n}{n!}\right]_{x_0}^{x}$$

$$m\frac{(x - x_0)^n}{n!} \leqslant R_n \leqslant M\frac{(x - x_0)^n}{n!}.$$

Now, M and m are the maximum and minimum values, respectively, of the

---

[2] We are assuming here that $x_0 < x$, so that the $(x - t)$ term is positive, or else that $n - 1$ is even, so that the numerator inside the integral is positive. If this is not the case, then the inequality signs should be reversed. This will not affect the outcome, however.

nth derivative $f^{(n)}(t)$ between $t = x_0$ and $t = x$. If this derivative is continuous, which is assumed, then somewhere between the values of M and m, say at $t = \xi$, must be a value $f^{(n)}(\xi)$ such that

$$R_n = f^{(n)}(\xi)\frac{(x - x_0)^n}{n!}, \quad \text{exactly.} \tag{2-8}$$

This is the remainder in the Lagrange form.

We now have the following theorem:

Taylor's Theorem *Let $f(x)$ be a function of x which is continuous, and whose first n derivatives are continuous, in the region between x_0 and x. Then $f(x)$ is given by*

$$f(x) = f(x_0) + f'(x_0)(x - x_0) + f''(x_0)\frac{(x - x_0)^2}{2!} + \cdots$$

$$+ f^{(n-1)}(x_0)\frac{(x - x_0)^{n-1}}{(n - 1)!} + R_n,$$

where R_n is as given in (2-8), and ξ lies between x_0 and x.

This expression can be made to apply to the Maclaurin series by letting $x_0 = 0$.

Unfortunately, this is not much help in calculating the value of the remaining terms, since it merely tells us that ξ is between x_0 and x, but does not tell us exactly which value to use for it. Since we do not know the value of ξ, we cannot calculate the nth derivative $f^n(\xi)$, and so cannot calculate R_n. And so we are almost back at the beginning—we still need all the terms of the infinite series to calculate $f(x)$ exactly.

On the other hand, if we are content *not* to calculate $f(x)$ exactly but to accept a small error, then we can use the expression for R_n to tell us the *maximum* error we might make by truncating the series.

We know that

$$R_n = f^{(n)}(\xi)\frac{(x - x_0)^n}{n!}. \tag{2-8}$$

Although we do not know the exact value of $f^{(n)}(\xi)$, we do know that ξ lies between x_0 and x, and so we can estimate its size; better than that, we can find its maximum size. Suppose this maximum is the value M:

$$|f^{(n)}(\xi)| \leqslant M \quad \text{for every } \xi \text{ between } x_0 \text{ and } x$$

Then

$$|R_n| \leqslant \left| M \frac{(x - x_0)^n}{n!} \right|.$$

The only requirement for this calculation is to find the maximum value of the nth derivative of $f(x)$ in the closed interval between x_0 and x. This is often quite easy, as we shall see in the following example.

2.4 Example 2-1. Finding the Value of the Exponential Constant e — Method I

The Maclaurin series for e^x is

$$e^x = 1 + x + \frac{x^2}{2!} + \frac{x^3}{3!} + \frac{x^4}{4!} + \frac{x^5}{5!} + \cdots.$$

Since we want the value of e, we merely let x equal 1 in the above series:

$$e^1 = 1 + 1 + \frac{1}{2!} + \frac{1}{3!} + \frac{1}{4!} + \frac{1}{5!} + \cdots$$

$$e = 1 + 1 + \frac{1}{2} + \frac{1}{6} + \frac{1}{24} + \frac{1}{120} + \cdots.$$

But since this is an infinite series, we must decide on the acceptable error to determine how many terms of the series are needed. If we decide that we want the final error to be less than 0.000001 we can then write

$$e = \underbrace{1 + 1 + \frac{1}{2} + \frac{1}{6} + \frac{1}{24} + \frac{1}{120} + \cdots + \frac{1}{(n-1)!}}_{n \text{ terms}} + R_n,$$

with the condition that

$$|R_n| \leqslant \left| M \frac{x^n}{n!} \right| = \left| M \frac{1}{n!} \right| \leqslant 0.000001.$$

Our aim is now to find the value of n that satisfies this last inequality; to do this, we need the value M, the maximum value of the derivative $f^{(n)}(\xi)$ for some ξ between x_0 (which is 0) and x (which is 1).

Since $f(x)$ is the function e^x,

$$f(x) = e^x$$
$$f^{(n)}(x) = e^x \quad \text{for any } n.$$

In the interval from 0 to 1, e^x takes on its maximum value when $x = 1$, so that the maximum derivative M is equal to e.

But since we are trying to find the exact value of e, it would be circular reasoning to use that value in finding M. Instead, let us pretend we do not know the exact value of e, but we do know that it is less than 3. This then gives us a maximum value

$$\left| f^{(n)}(\xi) \right| \leqslant M = 3.$$

We now return to our condition that

$$|R_n| \leqslant \left| M \frac{1}{n!} \right| \leqslant 0.000001.$$

Since M equals 3, and $n!$ is positive,

$$3\frac{1}{n!} \leqslant 0.000001.$$

We therefore need an n large enough so that

$$n! \geqslant 3 \times 10^6.$$

By trial and error, we now find that 9! is 362,880 and is therefore too small, while 10! is 3,628,800 and fulfils the requirement of being larger than 3×10^6. We therefore need 10 terms of the series to be assured of a truncation error less than 0.000001, although this is no guarantee that the answer as calculated by the computer will be this accurate; due to roundoff errors we may expect a greater error.

The tenth term of the series is $x^9/9!$, and so the actual computer program is as follows:

```
C      PROGRAM TO CALCULATE E TO THE X
       X = 1.
       E = 1 + X + X**2/2. + X**3/6. + X**4/24. + X**5/120. + X**6/720.
     1     + X**7/(7. * 720.)  + X**8 / (8. * 7. * 720.)
     2     + X**9 / (9. * 8. * 7. * 720.)
       WRITE (3, 10)  E
    10 FORMAT (' THE COMPUTER SAYS THAT E EQUALS ', F12.9)
       CALL EXIT
       END
```

The computer outputs the result

THE COMPUTER SAYS THAT E EQUALS 2.718280797

As expected, the answer of 2.718280797 is slightly under the correct value of 2.718281828 due to roundoff error. The roundoff error could have been decreased by adding the terms backward, or through a rather ingenious trick known as *Horner's rule*. In general, suppose that a polynomial in x is written as

$$y = a + bx + cx^2 + dx^3.$$

To evaluate this polynomial will require six multiplications and three additions (assuming that all the powers of x are calculated separately). With some factoring, we can reduce the equation as follows:

$$
\begin{aligned}
y &= a + bx + cx^2 + dx^3 \\
&= a + x[b + cx + dx^2] \\
&= a + x[b + x(c + dx)].
\end{aligned}
$$

The new form requires only three multiplications and three additions. Removing three multiplications removes three additional sources of roundoff error and speeds up the program.

This type of reduction can be performed on any polynomial in x, and provides even more dramatic savings of time and effort on large polynomials. Specifically, the first ten terms of the Maclaurin series

$$e^x = 1 + x + \frac{x^2}{2!} + \frac{x^3}{3!} + \frac{x^4}{4!} + \cdots + \frac{x^9}{9!} + R_n$$

form a polynomial in x, except that x equals 1 in our example. Using Horner's rule, we can rewrite the first ten terms as

$$e = 1 + (1 + \tfrac{1}{2}(1 + \tfrac{1}{3}(1 + \tfrac{1}{4}(1 + \tfrac{1}{5}(1 + \tfrac{1}{6}(1 + \tfrac{1}{7}(1 + \tfrac{1}{8}(1 + \tfrac{1}{9})))))))).$$

This equation is best translated into the FORTRAN statement

```
E = (((((((1. + 1./9.) /8. + 1.) /7. + 1.) /6. + 1.) /5.
    + 1.) /4. + 1.) /3. + 1.) /2. + 1.) + 1.
```

This statement performs the same calculation as the corresponding statement in the previous program, but has the advantage that it avoids addition of a large and a small number, which would lead to a roundoff problem.

The modified program is now:

```
C      REVISED PROGRAM TO CALCULATE E TO THE X
       E = ((((((((1. + 1./9.)/8. + 1.)/7. + 1.) / 6. + 1.) / 5. + 1.)
      1      / 4. + 1.) / 3. + 1.) / 2. + 1.) + 1.
       WRITE (3, 10)  E
   10 FORMAT (' THE COMPUTER SAYS THAT E EQUALS ', F12.9)
       CALL EXIT
       END
```

The new result shows a better agreement with theory:

THE COMPUTER SAYS THAT E EQUALS 2.718281751

The error between this answer and the correct value of 2.718281828 is only 0.000000077 (7.7×10^{-8}), far better than we had hoped. The reason is that at every step of the calculation the result of each addition or division is in the range of 0 to 3; since there are no abrupt changes of scale we do not run into serious roundoff.

The calculation of the remainder term is somewhat lengthy, and depends on the value of x we are using. For large values of x we shall obviously need more terms of the series, since the later terms have higher powers of x and therefore are larger; they contribute more to the sum. For this reason it is often useful to have the computer automatically compute the remainder term (if it can be easily expressed in terms of an explicit equation) and take only as many terms as are actually needed for the desired accuracy with the value of x used.

2.5 Computer Calculation of the Number of Terms Needed

Calculation of the remainder term R_n as such involves the determining of the maximum value of the nth derivative of the required function between x_0 and x. This requires some insight and often cannot be easily done by computer. But there are other methods of estimating the number of terms needed by the series for a given accuracy.

For an alternating series, the Leibnitz theorem we have already used is the most convenient. Briefly, it states that if a series (1) is strictly alternating in sign, (2) has each term smaller in magnitude than the previous term, and (3) the terms approach zero as a limit, then the series converges to a finite sum, and more important for us now, the error made by using only the first n terms is between zero and the first neglected term.

If a series satisfies the three conditions of this theorem, then we simply write the program so that terms are added until a term is encountered which is smaller than the required accuracy.

Very often the series will not satisfy the conditions of the theorem in its entirety, and yet the theorem can still be applied. For example, consider the sine series

$$\sin x = x - \frac{x^3}{3!} + \frac{x^5}{5!} - \frac{x^7}{7!} + \frac{x^9}{9!} - \cdots.$$

It is a strictly alternating series and, if we take enough terms, these terms approach zero as a limit. Yet it is not true that each term is smaller in magnitude than the previous term. As an example, suppose $x = 5$; then

$$\sin 5 = 5 - \frac{125}{6} + \frac{3125}{120} - \frac{78125}{5040} + \frac{1953125}{362880} - \cdots,$$

which is approximately

$$\sin 5 = 5 - 20.8 + 26 - 15.5 + 5.4 - \cdots.$$

For a time the terms get larger and larger since the numerator grows faster than the denominator for small factorials. Eventually, however, when $n > x$, $n!$ grows faster than x^n, and from then on the terms decrease in magnitude. Thus the conditions of Leibnitz' theorem are satisfied *provided* we wait until the terms start decreasing in magnitude before we truncate the series. If the series converges at all, then there must be such a point beyond which the terms decrease.

When a series does not alternate sign, then the procedure is not as simple, and we must actually consider the R_n term directly. Sometimes even here some simplification can be made if we know that $f^{(n)}(\xi)$ assumes its maximum value at the ends of the interval between x_0 and x. For example, in the series for e^x, if we take the first n terms, then the first neglected term will be

$$\frac{x^n}{n!}.$$

On the other hand, after the first n terms R_n is

$$R_n = f^{(n)}(\xi)\frac{x^n}{n!}.$$

If we know something about $f^{(n)}(\xi)$, such that it must be less than 3 in magnitude, then we know that the maximum error must be at most three times the first neglected term. We can often examine R_n and determine some bound for $f^{(n)}$. It may be that $f^{(n)}$ in the range is such that the maximum error, as obtained from R_n, is of the same order of magnitude as the first neglected term. In that case, the situation becomes similar to that of a convergent alternating series—we merely take terms until we reach a term which is smaller than the maximum acceptable error, and then stop.

2.6 Example 2-2. Finding the Value of the Exponential Constant *e* — Method II

In order to save the effort of evaluating R_n by hand to determine the number of terms needed, we rewrite the program of Example 2-1 to permit automatic calculation of terms until a term smaller than the desired error (in this case 0.000001) is reached.

In this program a one-dimensional array called TERM stores successive terms found in the first part of the program (until statement 100). An IF statement stops this part of the program when a term is reached which is less than 0.000001 in magnitude. Since all the terms are positive in this case, the IF test is simple; if the series were alternating, a more complex test would be needed to test for the magnitude only.

When all the terms of the series are calculated, the variable I holds the number of terms required. These I terms are then separately summed, once forward and once backward, as a test of roundoff error.

The following program is used.

```
C       PROGRAM TO CALCULATE E TO THE X FORWARD AND BACKWARD
        DIMENSION TERM (50)
C       COLLECT ALL THE TERMS
        FACT = 1.
        DO 100 I = 1, 50
        FACT = FACT * I
        TERM (I) = 1. / FACT
        IF (TERM (I) - .000001) 200, 100, 100
    100 CONTINUE
C       ALL THE TERMS HAVE BEEN CALCULATED, AND I HOLDS THE NUMBER OF
C       TERMS NEEDED FOR THE LAST TERM TO BE LESS THAN .000001
    200 WRITE (3, 210) I
    210 FORMAT (' WE NEED ', I5, ' TERMS.')
        SUM = 1.
        DO 300 J = 1, I
    300 SUM = SUM + TERM (J)
        WRITE (3, 310) SUM
    310 FORMAT (' SUMMING FORWARD WE GET THE ANSWER ',F15.8)
        SUM = 0.
        DO 400 J = 1, I
        K = I + 1 - J
    400 SUM = SUM + TERM (K)
        SUM = SUM + 1.
        WRITE (3, 410) SUM
    410 FORMAT(' SUMMING BACKWARD WE GET THE ANSWER',  F15.8)
        CALL EXIT
        END
```

As seen from the following printout, ten terms of the series are needed:

```
WE NEED     10 TERMS.
SUMMING FORWARD WE GET THE ANSWER     2.71828079
SUMMING BACKWARD WE GET THE ANSWER    2.71828222
```

Summing forward, a roundoff error is still made; summing backward, however, the result is 2.71828222, which is within 0.0000004 of the correct answer of 2.718281828.

This accuracy is about all we can expect without going to double precision arithmetic since we cannot apply Horner's rule as before if we do not know the number of needed terms.

It is interesting to examine the problem of finding $e^{10.4}$, for example, instead of e^1. Whereas we needed only 10 terms for e^1, we would need hundreds of terms to find $e^{10.4}$. In the process, we would find that we cannot calculate $n!$ or even x^n when n becomes large enough, because of machine overflow. In order to permit calculation of the terms, we would have to alternately multiply each term by x and divide by the next part of the factorial so as to try to counterbalance a large numerator with a large denominator. Fortunately, we do not have to bother with this, since

$$e^{10.4} = e^{10} \, e^{0.4}.$$

Only a few terms of the series are needed to find $e^{0.4}$, while e^{10} could be found by multiplying e by itself ten times. The roundoff error on this multiplication could get serious, however, and so we might either have to use some special techniques to perform the multiplication, or else find powers of e separately and store them in a small table in memory.

2.7 Example 2-3. Calculating the Sine

As a further example, suppose we wish to write a generalized program to find the sine of *any* angle. How many terms of the series do we need?

If the series is to work for *any* angle, no matter how large, we are tempted to let the computer calculate terms until it finds a term smaller than the required error. Since the sine series is an alternating one, this would guarantee that the overall truncation error is less than the first neglected term. Although this procedure may work (except for roundoff error) it will take a very long time for large angles, where very many terms will be required.

An alternative is to note that the sine function repeats itself every 2π radians. By subtracting a multiple of 2π from large angles, they can be converted to an angle between 0 and 2π. By a further slight change we now change all angles to lie between $-\pi/2$ and $+\pi/2$ as follows:

Angles between	Change	New angle between
0 and $\pi/2$	No change	0 and $\pi/2$
$\pi/2$ and $3\pi/2$	Subtract angle from π	$\pi/2$ and $-\pi/2$
$3\pi/2$ and 2π	Subtract 2π from angle	$-\pi/2$ and 0

Negative angles are converted in the same way.

The problem is now reduced to finding the sine of an angle between $-\pi/2$ and $+\pi/2$ radians. Using the sine series

$$\sin x = x - \frac{x^3}{3!} + \frac{x^5}{5!} - \frac{x^7}{7!} + \frac{x^9}{9!} - \cdots$$

we see that x will vary only from $-\pi/2$ (or -1.5708) to $+1.5708$ and so we shall need fewer terms of the series. In order to decide on the number of terms needed let us arbitrarily decide that an accuracy of 0.000005 will do for our application. We then look at the remainder term from Taylor's formula

$$R_n = f^{(n)}(\xi)\frac{x^n}{n!}$$

$$|R_n| \leq \left| M\frac{x^n}{n!} \right|,$$

where M is the maximum value of the nth derivative of the sine. Since the derivatives of the sine are $\cos x$, $-\sin x$, $-\cos x$, $\sin x \ldots$ the maximum value of all derivatives for any x is always 1. Further, for an accuracy of 0.000005 R_n cannot be larger than this value, so that

$$|R_n| \leq \left| \frac{x^n}{n!} \right| \leq 0.000005,$$

where x may be as large as ± 1.5708 radians. Assuming the positive value, we can remove the absolute value signs and write

$$\frac{x^n}{n!} \leq 0.000005 = 5 \times 10^{-6}.$$

By trial and error we find that at the maximum $x = 1.5708(\pi/2$ radians or $90°)$

$$\frac{x^{11}}{11!} = 3.6 \times 10^{-6},$$

so that $n = 11$ terms are required.

But this is a bit misleading in this case, since some of the terms of the sine series are zero. By definition, R_n was the error obtained by using the first n terms, and omitting the terms starting with $f^{(n)}(x)/(x^n/n!)$. Some of the first n terms are zero, so that when we take all terms up to, but not including, $x^{11}/11!$ we are really taking 11 terms although we need only calculate the five nonzero terms.

As a test of the series, we can write a short test program to print a table of sines up to 90 degrees, and compare the series results with the FORTRAN sine function.[3] For minimum roundoff we calculate the small terms first:

```
      WRITE (3, 10)
10 FORMAT (' DEGREES      RADIANS     COMPUTED SINE     ACTUAL SINE
   1  ERROR')
      DO 100 I = 10, 90, 10
      X = I / 57.2957795
      S1 = (X**9 / 362880.) - (X**7 / 5040.) + X**5 / 120. - X**3/6. + X
      S2 = SIN (X)
      ERROR = S1 - S2
      WRITE (3, 50) I, X, S1, S2, ERROR
50 FORMAT (' ', I4, 7X, F9.7, 5X, F9.7, 7X, F9.7, 3X, E13.6)
100 CONTINUE
      CALL EXIT
      END
```

The following output shows that the maximum error is only 3.54×10^{-6}, slightly below the calculated maximum and within the desired accuracy:

DEGREES	RADIANS	COMPUTED SINE	ACTUAL SINE	ERROR
10	0.1745329	0.1736481	0.1736481	0.116415E-09
20	0.3490658	0.3420201	0.3420201	0.000000E 00
30	0.5235987	0.5000000	0.5000000	0.000000E 00
40	0.6981317	0.6427876	0.6427876	0.931322E-09
50	0.8726646	0.7660444	0.7660444	0.558793E-08
60	1.0471975	0.8660254	0.8660254	0.419095E-07
70	1.2217304	0.9396928	0.9396926	0.224448E-06
80	1.3962634	0.9848087	0.9848077	0.973232E-06
90	1.5707963	1.0000035	1.0000000	0.354088E-05

2.8 Improving on the Taylor Series

Theoretically, the Taylor (or Maclaurin) series can provide perfect accuracy *if* we take an infinite number of terms. Since this is impractical, we must limit the number of terms we take to some finite number, which is a compromise between speed and accuracy. The question arises: can we change the series to get better accuracy with fewer terms? Perhaps.

Looking for the above results, we see that the sine computed from the series is quite accurate for small angles, and that the error increases quite quickly for angles near 90 degrees. We can see this quite clearly in curve A in Figure 2-3, which shows that the error below 50 degrees is extremely small, and rises very steeply near 90 degrees.

Our object is to reduce the maximum error near 90 degrees by somehow changing the series to spread out the error over a larger range of angles. One such possible change is to reduce the coefficient of the x^9 term since we suspect that it is this term which is causing a large part of the error.

---

[3] The sample program was run in extended precision, and the FORTRAN SIN function itself is accurate to about 10^{-9}. Hence errors in the table of 10^{-9} could be due to the SIN as well.

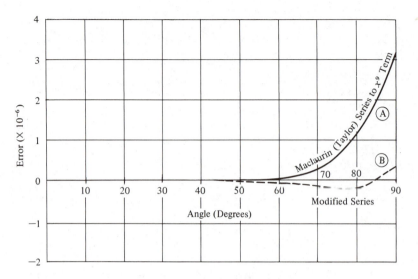

Figure 2-3.   Errors of two series for the sine.

The x^9 term is normally $x^9/9!$ or $x^9/362880$. We reduce this term slightly by increasing the denominator to 370500 in the following FORTRAN statement:

```
S1 = (X**9 / 370500.) - (X**7 / 5040.) + X**5 / 120. - X**3/6. + X
```

Rerunning the program we obtain the following results:

DEGREES	RADIANS	COMPUTED SINE	ACTUAL SINE	ERROR
10	0.1745329	0.1736481	0.1736481	0.116415E-09
20	0.3490658	0.3420201	0.3420201	0.000000E 00
30	0.5235987	0.5000000	0.5000000	-0.465661E-09
40	0.6981317	0.6427876	0.6427876	-0.139698E-08
50	0.8726646	0.7660444	0.7660444	-0.111758E-07
60	1.0471975	0.8660253	0.8660254	-0.447034E-07
70	1.2217304	0.9396925	0.9396926	-0.119209E-06
80	1.3962634	0.9848075	0.9848077	-0.170432E-06
90	1.5707963	1.0000002	1.0000000	0.241212E-06

Replotting the error, we get curve B in Figure 2-3. If we examine the curves and the two error printouts, we see that we have not reduced the error; we have merely spread it out over a larger area, and reduced the *maximum* error. Although Figure 2-3 does not show the increased error below 50 degrees, a comparison of the two printouts shows that the error here has increased.

We are so far proceeding more by intuition than by mathematical logic, but we suspect that we could reduce the maximum error even more by adjusting some of the other terms of the original series. Rather than use a trial-and-error approach, however, we would like to develop some mathematical strategy on how to do this.

In order to develop a system, we wish to normalize the series to make it more universal. Specifically, our Maclaurin series for the sine has so far been used for angles from -90 degrees to $+90$ degrees, or to be more exact, for angles from $-\pi/2$ and $+\pi/2$ radians. Since this restricts us to the sine series, we want to generalize by having the independent variable in the series go from -1 to $+1$, instead of from -1.5708 to $+1.5708$. We do this by a simple change of variable; if x is the original angle in radians, let

$$x' = \frac{x}{\pi/2} = \frac{2x}{\pi}$$

$$x = \frac{\pi x'}{2}.$$

If x' now goes from -1 to $+1$, x goes from $-\pi/2$ to $+\pi/2$, as required.
   Using x' instead of x, we can now rewrite the original sine series as

$$\sin x = x - \frac{x^3}{3!} + \frac{x^5}{5!} - \frac{x^7}{7!} + \frac{x^9}{9!} - \cdots$$

$$= \left(\frac{\pi x'}{2}\right) - \frac{\left(\frac{\pi x'}{2}\right)^3}{3!} + \frac{\left(\frac{\pi x'}{2}\right)^5}{5!} - \frac{\left(\frac{\pi x'}{2}\right)^7}{7!} + \frac{\left(\frac{\pi x'}{2}\right)^9}{9!} - \cdots.$$

Let us now drop the primes, remembering from now on that the x is a normalized x running from -1 to $+1$; to get it we either divide the number of radians by $\pi/2$, or else we divide the number of degrees by 90.
   Simplifying each term of the series by multiplying out by the appropriate power of $\pi/2$ and dividing by the factorial, we get the series

$$\sin\left(\frac{\pi x}{2}\right) = 1.570796329x - 0.6459640960x^3 + 0.07969262599x^5$$

$$-0.004681754102x^7 + 0.0001604411842x^9.$$

   Before proceeding, we wish to check our arithmetic, and so we run the following program as a check:

```
      WRITE (3, 10)
   10 FORMAT (' DEGREES       RADIANS      COMPUTED SINE      ACTUAL SINE
      1  ERROR')
      DO 100 I = 10, 90, 10
      R = I / 57.2957795
      S = SIN (R)
      X = I / 90.
      TAYLR = X * (1.570796329 - X * X * (0.6459640960 - X * X *
      1   (0.07969262599 - X * X * (0.004681754102 - X * X *
      2   (0.0001604411842)))))
      ERROR = TAYLR - S
```

```
100 WRITE (3,   50) I, R, TAYLR, S, ERROR
 50 FORMAT (' ',  I4,  7X,  F9.7,  5X,  F9.7,  7X,  F9.7,  3X,  E13.6)
    CALL EXIT
    END
```

The results show that this series is identical to the original Maclaurin series (except for different roundoff errors):

DEGREES	RADIANS	COMPUTED SINE	ACTUAL SINE	ERROR
10	0.1745329	0.1736481	0.1736481	0.465661E-09
20	0.3490658	0.3420201	0.3420201	0.465661E-09
30	0.5235987	0.5000000	0.5000000	0.465661E-09
40	0.6981317	0.6427876	0.6427876	0.186264E-08
50	0.8726646	0.7660444	0.7660444	0.651925E-08
60	1.0471975	0.8660254	0.8660254	0.428408E-07
70	1.2217304	0.9396928	0.9396926	0.226311E-06
80	1.3962634	0.9848087	0.9848077	0.976026E-06
90	1.5707963	1.0000035	1.0000000	0.354461E-05

Having now normalized our series and put it into a more universal form, we note that this can be done to any other series by a change of variable, and so we are not restricting ourselves merely to functions defined over the interval from -1 to $+1$.

2.9 Chebyshev Polynomials

We found earlier that terms up to, but not including, the x^{11} term of the standard Maclaurin (Taylor) series were necessary to get accuracy of 0.000005, but we also found that by some changing of the coefficients we could decrease the maximum error by spreading it out over a larger range of angles. We are now about to show how to do this.

We have two choices open to us. If we are satisfied with the accuracy obtained but wish to simplify the calculations, we can aim to get the same accuracy with fewer terms (such as only up to the x^7 term instead of the x^9 term); on the other hand, if we do not mind the calculations, we might keep terms up to x^9 but shoot for better accuracy. Let us assume that the accuracy of the original series using terms up to x^9 is good enough for us, but we want to shorten the calculations by eliminating the last term.

To do this we start by introducing the *Chebyshev polynomials of the first kind*. Their symbol is $T_n(x)$, and they are polynomials in x^n.

The Chebyshev polynomials are defined as follows:

$$T_n(x) = \cos (n \cos^{-1} x), \qquad (2\text{-}9)$$

where n can be any integer, although we generally consider only nonnegative integers (including 0).

If $n = 0$, then

$$T_0(x) = \cos (0 \cdot \cos^{-1} x) = \cos 0 = 1.$$

If $n = 1$, then

$$T_1(x) = \cos (1 \cdot \cos^{-1} x) = \cos (\cos^{-1} x) = x.$$

Finding $T_2(x)$ is somewhat harder, and so we use some mathematical sleight-of-hand; we use a change of variable, and let

$$\phi = \cos^{-1} x$$
$$x = \cos \phi.$$

The defining Eq. (2-9) then becomes

$$T_n(x) = \cos n\phi. \tag{2-10}$$

Similarly, we can write the following two equations:

$$T_{n+1}(x) = \cos [(n + 1)\phi] = \cos (n\phi + \phi) = \cos n\phi \cos \phi - \sin n\phi \sin \phi$$
$$T_{n-1}(x) = \cos [(n - 1)\phi] = \cos (n\phi - \phi) = \cos n\phi \cos \phi + \sin n\phi \sin \phi.$$

Adding the two equations we have part of the right side cancelling, so that

$$T_{n+1}(x) + T_{n-1}(x) = 2 \cos n\phi \cos \phi = 2 \cdot T_n(x) \cdot x,$$

since $\cos n\phi = T_n(x)$, and $\cos \phi = T_1(x) = x$.

Rearranging the last equation, we have

$$T_{n+1}(x) + T_{n-1}(x) = 2xT_n(x)$$
$$T_{n+1}(x) = 2xT_n(x) - T_{n-1}(x). \tag{2-11}$$

The latter is a recurrence relation which lets us find any Chebyshev polynomial if we know the preceding two. For example,

$$T_2(x) = 2xT_1(x) - T_0(x) = 2x \cdot x - 1 = 2x^2 - 1$$

Table 2-1 gives the first twelve Chebyshev polynomials.

The interesting properties of these polynomials are hinted at by the details in the table. For example, all the coefficients for any of the polynomials add to $+1$; this means that if $x = 1$, each of the polynomials equals 1, and so on —more on this in a moment.

TABLE 2-1

Chebyshev Polynomials of the First Kind

$T_0(x) = 1$
$T_1(x) = x$
$T_2(x) = 2x^2 - 1$
$T_3(x) = 4x^3 - 3x$
$T_4(x) = 8x^4 - 8x^2 + 1$
$T_5(x) = 16x^5 - 20x^3 + 5x$
$T_6(x) = 32x^6 - 48x^4 + 18x^2 - 1$
$T_7(x) = 64x^7 - 112x^5 + 56x^3 - 7x$
$T_8(x) = 128x^8 - 256x^6 + 160x^4 - 32x^2 + 1$
$T_9(x) = 256x^9 - 576x^7 + 432x^5 - 120x^3 + 9x$
$T_{10}(x) = 512x^{10} - 1280x^8 + 1120x^6 - 400x^4 + 50x^2 - 1$
$T_{11}(x) = 1024x^{11} - 2816x^9 + 2816x^7 - 1232x^5 + 220x^3 - 11x$

Just as the Chebyshev polynomials $T_n(x)$ can be written as polynomials in x, so we can solve backward and write x in terms of the $T_n(x)$. For example,

$$x^0 = 1 = T_0(x)$$
$$x^1 = x = T_1(x).$$

Further, since

$$T_2(x) = 2x^2 - 1,$$

we can solve for x^2:

$$2x^2 = T_2(x) + 1 = T_2(x) + T_0(x)$$
$$x^2 = \tfrac{1}{2}[T_2(x) + T_0(x)].$$

We can express any power of x as a linear combination of the $T_n(x)$ as in Table 2-2. Hence the Chebyshev polynomials can be expressed in terms of x, and conversely x can be expressed in terms of the Chebyshev polynomials. These expressions are identities, and actually hold for all x, even outside the range of $-1 \leqslant x \leqslant +1$.

But from the original definition

$$T_n(x) = \cos(n \cos^{-1} x)$$

we see that each of the T_n is only defined as the cosine of some quantity; hence it cannot be greater than ± 1. Similarly, the expression $\cos^{-1} x$ is only defined if x is within -1 and $+1$.

TABLE 2-2

Powers of x in Terms of Chebyshev Polynomials T_n

$x^0 = T_0$

$x^1 = T_1$

$x^2 = \dfrac{1}{2}(T_2 + T_0)$

$x^3 = \dfrac{1}{4}(T_3 + 3T_1)$

$x^4 = \dfrac{1}{8}(T_4 + 4T_2 + 3T_0)$

$x^5 = \dfrac{1}{16}(T_5 + 5T_3 + 10T_1)$

$x^6 = \dfrac{1}{32}(T_6 + 6T_4 + 15T_2 + 10T_0)$

$x^7 = \dfrac{1}{64}(T_7 + 7T_5 + 21T_3 + 35T_1)$

$x^8 = \dfrac{1}{128}(T_8 + 8T_6 + 28T_4 + 56T_2 + 35T_0)$

$x^9 = \dfrac{1}{256}(T_9 + 9T_7 + 36T_5 + 84T_3 + 126T_1)$

$x^{10} = \dfrac{1}{512}(T_{10} + 10T_8 + 45T_6 + 120T_4 + 210T_2 + 126T_0)$

$x^{11} = \dfrac{1}{1024}(T_{11} + 11T_9 + 55T_7 + 165T_5 + 330T_3 + 462T_1)$

Hence, although the conversions between x and $T_n(x)$ given in Tables 2-1 and 2-2 may tempt us to use any x and any T_n, we must remember that only values between -1 and $+1$ are allowed. If we plot the T_n as a function of the x, as in Figure 2-4, the entire plot must lie within the dashed square shown since the function is not defined outside regardless of what Table 2-1 may imply.

What makes the Chebyshev polynomials so interesting is their behavior in the region from $x = -1$ to $x = +1$, which we see in Figure 2-4.

Only the first four polynomials are shown in Figure 2-4, but all the polynomials share certain common features:

1. In the region $-1 \leqslant x \leqslant +1$ the magnitude of each T_n is less than or equal to 1, so that all the polynomials stay within the dashed square shown in Figure 2-4.
2. The coefficient of the highest power of x in each polynomial is 2^{n-1}; thus the coefficient of the highest power in T_5 is 2^4 or 16. If we leave this coefficient unchanged, but try changing *any* other coefficient in $T_n(x)$, we

find that the graph of that Chebyshev polynomial no longer stays within the dashed square.

This is often called the *minimax* property and is our prime reason for using the Chebyshev polynomials at all. Rephrased, the minimax property says that, of all the polynomials in x of order n having the same leading

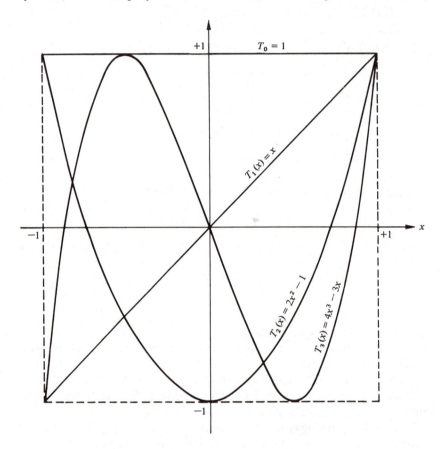

Figure 2-4. Chebyshev polynomials T_0 through T_3.

coefficient as T_n, the Chebyshev polynomial is the only one whose maximum value in the region $-1 \leqslant x \leqslant +1$ is less than or equal to 1. Hence it has the *mini*mum *maxi*mum value. Changing any one of the coefficients (other than the first) or any combination of them, even by the slightest amount, will always raise one of the peaks or lower one of the valleys, and bring the function outside the square.

3. If we set $T_n(x) = 0$, we may solve for the roots of the polynomial; that

is, those values of x which satisfy this equation. At these value of x the graph of the polynomial crosses the x axis. Each of the polynomials has exactly n real roots, and therefore crosses the x axis in n places. We can see this in either Figure 2-4 or Figure 2-5. Between these roots the polynomial swings to $+1$ or -1.

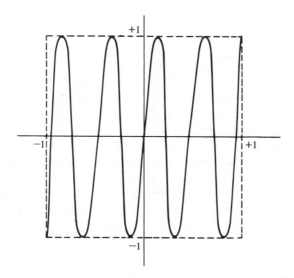

Figure 2-5.   Plot of $T_9(x)$ in the region from $x = -1$ to $x = +1$.

2.10 Telescoping a Series

We are now ready to try to shorten a series; this process is called *telescoping*. As you remember, the purpose is to reduce the amount of calculations needed by the computer, yet retain the same accuracy, if possible.

Our procedure will be as follows. We start with a power series, such as the series for a sine, exponential, or any other function, which may be generalized as

$$f(x) = a_0 + a_1 x + a_2 x^2 + a_3 x^3 + a_4 x^4 + \cdots.$$

By analyzing the remainder R_n we decide that n terms are sufficient for the needed accuracy, and use a truncated series

$$f(x) = a_0 + a_1 x + a_2 x^2 + \cdots + a_{n-2} x^{n-2} + a_{n-1} x^{n-1}.$$

By referring to Table 2-2, we express each power of x in terms of the various

T_n and simplify. We then get a new series in terms of the Chebyshev polynomials, with new coefficients:

$$f(x) = b_0 + b_1 T_1 + b_2 T_2 + \cdots + b_{n-2} T_{n-2} + b_{n-1} T_{n-1}.$$

As it turns out,

$$b_{n-1} = \frac{a_{n-1}}{2^{n-2}} \qquad \text{(see Table 2-2)},$$

so that the coefficient of the highest Chebyshev polynomial is much smaller than the coefficient of the original highest power of x.

In the original Taylor series, the highest term in x was $a_{n-1} x^{n-1}$, where x had been normalized to lie between -1 and $+1$. If we had decided to truncate this term then we would have created an error, depending on the value of x, as high as $\pm a_{n-1}$ in magnitude; this error would have been in addition to the truncation error produced by dropping the higher terms beyond the nth, and could have made a rather large difference in the final result.

In the new series with Chebyshev polynomials, the last term is

$$b_{n-1} T_{n-1} = \frac{a_{n-1}}{2^{n-2}} T_{n-1},$$

where the T_{n-1} again has a maximum value of ± 1, except that it reaches this value several times over the range. If we now truncate this term we introduce an additional error as high as

$$\pm \frac{a_{n-1}}{2^{n-2}}.$$

This error is smaller by a factor of 2^{n-2} than the error which would have been introduced by truncating the highest term in the original series. But there is a price—since the T_{n-1} reaches its maximum value of ± 1 several times (to be specific, $n - 2$ times) throughout the interval, we make this new error for many different values of x, rather than just when x is near ± 1. We are essentially taking a large error and spreading it out over a wide range of x's, making its maximum value smaller.

Having truncated the last term, we have a new series

$$f(x) = b_0 + b_1 T_1 + b_2 T_2 + \cdots + b_{n-2} T_{n-2}.$$

As a last step, we now refer to Table 2-1 and replace each T_n by the appropriate powers of x, and then simplify. This gives a new series in x, with new coefficients:

$$f(x) = c_0 + c_1 x + c_2 x^2 + \cdots + c_{n-2} x^{n-2},$$

which has only $(n - 1)$ terms, one less than the original series.

The technique is illustrated in the following detailed example.

2.11 Example 2-4. Telescoping the Sine Series

We found earlier that the normalized sine series was

$$\sin\left(\frac{\pi x}{2}\right) = 1.570796329x - 0.6459640960x^3 + 0.07969262599x^5$$

$$-0.004681754102x^7 + 0.0001604411842x^9.$$

Referring to Table 2-2 we substitute the Chebyshev polynomials T_1 through T_9 for the powers of x as follows:

$$\begin{aligned}
\sin\left(\frac{\pi x}{2}\right) = {}& 1.570796329T_1 \\
& - 0.6459640960[\tfrac{1}{4}(T_3 + 3T_1)] \\
& + 0.07969262599[\tfrac{1}{16}(T_5 + 5T_3 + 10T_1)] \\
& - 0.004681754102[\tfrac{1}{64}(T_7 + 7T_5 + 21T_3 + 35T_1)] \\
& + 0.0001604411842[\tfrac{1}{256}(T_9 + 9T_7 + 36T_5 + 84T_3 + 126T_1)].
\end{aligned}$$

We now simplify and combine like polynomials together:

$$\begin{aligned}
\sin\left(\frac{\pi x}{2}\right) = {}& 1.133649778T_1 - 0.1380706342T_3 \\
& + 0.004491284309T_5 - 0.00006751189749T_7 \\
& + 0.0000006267233758T_9.
\end{aligned}$$

To telescope the series, we now propose to cut off the T_9 term, and should seriously examine the error this will cause.

So far, the substitution of the Chebyshev polynomials has not introduced any new error (except for possible roundoff in calculating the new coefficients), and the original truncation error introduced by truncating from the x^{11} term on is still the only error; the introduction of Chebyshev polynomials can be looked at as a simple change of variable.

To see the error introduced by dropping the T_9 term, we refer to a graph of the T_9 polynomial, Figure 2-5. When we cut off the T_9 term we make an *additional* error (over and above the original truncation error) which can be as large as the size of the term cut off, namely

$$E_{\text{telescoping}} = 0.0000006267233758 T_9.$$

But as we see from Figure 2-5, T_9 never exceeds 1 in magnitude:

$$|T_9| \leqslant 1$$
$$|E_{\text{telescoping}}| \leqslant 0.0000006267233758 \cdot 1$$
$$\leqslant 6.3 \times 10^{-7} = 0.63 \times 10^{-6}.$$

Now, the original truncation error can be either positive or negative, and the telescoping error can be either positive or negative (depending on the value of x), and so the total error, being the sum of the two errors, can be positive, negative, or even zero, depending on whether the two errors add to each other, or whether (if the signs are opposite) they actually subtract from each other.

The only conclusion we can draw at this point concerns the *maximum* total error, which we can get from the triangle inequality:

$$|E_{\text{total}}| \leqslant |E_{\text{truncation}}| + |E_{\text{telescoping}}|$$
$$\leqslant 3.6 \times 10^{-6} + 0.63 \times 10^{-6}$$
$$\leqslant 4.23 \times 10^{-6}.$$

Thus the process of telescoping increases the maximum error only slightly and may even decrease it, since the two errors might subtract from each other.

But we must still convert back to powers of x. After we remove the T_9 term, we are left with

$$\sin\left(\frac{\pi x}{2}\right) = 1.133649778 T_1 - 0.1380706342 T_3$$
$$+ 0.004491284309 T_5 - 0.00006751189749 T_7,$$

and we now convert back to a polynomial in x by referring to Table 2-1 and substituting for each of the Chebyshev polynomials:

$$\sin\left(\frac{\pi x}{2}\right) = 1.133649778 x$$
$$- 0.1380706342(4x^3 - 3x)$$
$$+ 0.004491284309(16x^5 - 20x^3 + 5x)$$
$$- 0.00006751189749(64x^7 - 112x^5 + 56x^3 - 7x).$$

We again simplify and collect like powers of x together to get a telescoped power series with terms up to only x^7:

$$\sin\left(\frac{\pi x}{2}\right) = 1.570790685x - 0.6458888889x^3 + 0.07942188149x^5$$
$$- 0.004320761439x^7.$$

The coefficients of this power series are quite similar to the coefficients of the original normalized Maclaurin series, but are just slightly different.

As a test of the telescoped series, we may run the following program to print a table of sines and compare with the normal FORTRAN function:

```
      WRITE (3, 10)
 10 FORMAT (' DEGREES      RADIANS      COMPUTED SINE     ACTUAL SINE
    1   ERROR')
      DO 100 I = 10, 90, 10
      R = I / 57.2957795
      S = SIN (R)
      X = I / 90.
      TELES = X * (1.570790685 - X * X * (0.6458888889 - X * X *
    1   (0.07942188149 - X * X * (0.004320761439))))
      ERROR = TELES - S
100 WRITE (3,  50) I, R, TELES, S, ERROR
 50 FORMAT (' ', I4, 7X, F9.7, 5X, F9.7, 7X, F9.7, 3X, E13.6)
      CALL EXIT
      END
```

The program gives us the following results:

DEGREES	RADIANS	COMPUTED SINE	ACTUAL SINE	ERROR
10	0.1745329	0.1736476	0.1736481	-0.528059E-06
20	0.3490658	0.3420195	0.3420201	-0.565778E-06
30	0.5235987	0.4999999	0.5000000	-0.526197E-07
40	0.6981317	0.6427881	0.6427876	0.528991E-06
50	0.8726646	0.7660449	0.7660444	0.525265E-06
60	1.0471975	0.8660252	0.8660254	-0.135041E-06
70	1.2217304	0.9396922	0.9396926	-0.393949E-06
80	1.3962634	0.9848089	0.9848077	0.123400E-05
90	1.5707963	1.0000029	1.0000000	0.291503E-05

Figure 2-6 gives us a comparison of the original series having terms up to x^9, the telescoped series obtained from this by removing the x^9 term, and a telescoped series obtained from a series going to x^{11} by removing the last term.

As compared with the original series (curve A), we can see that we have a real choice—either accept the same maximum error (though spread out over the entire interval) by reducing the number of terms by telescoping so that the highest term is only an x^7 instead of an x^9 term (Curve B), or else start from a higher order approximation and telescope from a series with an x^{11} term down to a series having only an x^9 term (Curve C).

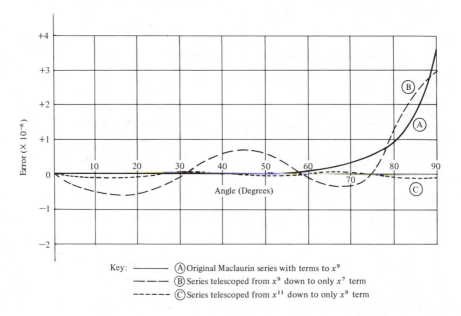

Key: ———— Ⓐ Original Maclaurin series with terms to x^9
     – – – – Ⓑ Series telescoped from x^9 down to only x^7 term
     – – – – – Ⓒ Series telescoped from x^{11} down to only x^9 term

Figure 2-6.   Error curve for three different sine series.

2.12 Conclusions

It may seem trivial to spend so much time calculating coefficients and making conversions to and from Chebyshev polynomials just to cut one term from a series. But there is a compromise to be made between speed and accuracy in any program, and in a program which is used very much, or which is repeated very many times, a small saving in time (and, incidentally, also in computer memory space required) may be very important in the long run.

2.13 Problems

1. Write the Maclaurin series for $f(x) = 3x^3 - 5x^2 + x - 2$.

2. Expand $f(x) = 3x^3 - 5x^2 + x - 2$ in a Taylor series about $x_0 = 1$, and about $x_0 = 3$, and simplify; compare with the results of Problem 1.

3. Write the Maclaurin expansions for the following functions for the first five terms, and write the expression for R_5:

a. $\ln(x+1)$
b. $\tan x$
c. $\sin^{-1} x$
d. $\tan^{-1} x$
e. $\sinh x$

4. Expand $f(x) = e^x \sin x$ in a Maclaurin series. Now take the series for the sine and the series for e^x and multiply as follows, and compare the results:

$$x \quad\quad -\frac{x^3}{3!} \quad\quad +\frac{x^5}{5!} \quad\quad -\frac{x^7}{7!} \quad\quad +\cdots$$

$$\times\left[1 + x + \frac{x^2}{2!} + \frac{x^3}{3!} + \frac{x^4}{4!} + \frac{x^5}{5!} + \frac{x^6}{6!} + \frac{x^7}{7!} + \cdots\right.$$

$$x \quad\quad -\frac{x^3}{3!} \quad\quad +\frac{x^5}{5!} \quad\quad -\frac{x^7}{7!} \quad\quad \cdots$$

$$+\, x^2 \quad\quad -\frac{x^4}{3!} \quad\quad +\frac{x^6}{5!} \quad\quad -\frac{x^8}{8!}$$

$$\frac{x^3}{2!} \quad \cdots$$

$$\vdots$$

$$x + x^2 + x^3\left\{\frac{1}{2!} - \frac{1}{3!}\right\} + \cdots.$$

5. Write the Taylor series for $\sin x$, expanded about $x_0 = \pi/2$.

6. Write the Taylor series for $f(x) = \sqrt{x}$, expanded about $x_0 = 1$. Can this function be expanded in a Maclaurin series? Why?

7. How many terms are needed to evaluate each of the following to within ± 0.001:

a. $\sin x$ for x between 0 and $\pi/2$ radians, using Maclaurin series.
b. $\cos x$ for x between 0 and $\pi/2$ radians, using Maclaurin series.
c. e^x for x between 0 and 2, using a Maclaurin series.

d. $\sqrt{x}$ for x between 0.5 and 1.5, using the Taylor series of problem 6.

8. Write the series for e^{3x} as a Maclaurin series; now take the series for e^x and substitute $3x$ for every occurrence of x, and compare.

9. Using $j = \sqrt{-1}$, we can write

$$\cos \theta = \tfrac{1}{2}(e^{j\theta} + e^{-j\theta})$$

By substituting $j\theta$ and $-j\theta$ into the e^x series, obtain the two Maclaurin series for $e^{j\theta}$ and for $e^{-j\theta}$. Add them and divide by 2, and compare with the series for $\cos \theta$.

10. Write the Taylor series for $\ln x$ by expanding about $x_0 = 2$. What is the region of convergence?

11. If we used the Maclaurin series for $\cos x$ to determine $\cos 100$ radians (an extremely bad practice, by the way), how many terms would we have to take before the Leibnitz theorem would apply?

12. Find the roots of $T_3(x)$ (the points where T_3 crosses the x axis), and the values of x where T_3 has its maximum and minimum values.

13. The Chebyshev polynomials have an *orthogonality* property which makes them of great interest. It is expressed by the following integral:

$$\int_{-1}^{+1} \frac{T_m(x)T_n(x)}{\sqrt{1 - x^2}}\,dx \quad \begin{cases} = 0 \text{ whenever } m \neq n \\[2mm] = \pi \text{ whenever } m = n = 0 \\[2mm] = \frac{\pi}{2} \text{ whenever } m = n = 1, 2, \ldots. \end{cases}$$

Show that this is true.

14. Write the first three terms of the Maclaurin series for e^x, and telescope down to two terms. Compare with the two-term Maclaurin series. Graph the function e^x for $-1 \leqslant x \leqslant +1$, and compare with the Taylor series of two terms and the telescoped series. What is the maximum error in the two series approximations?

15. Using the Maclaurin series of problem 1, telescope to only 3 terms, and graph the original function and the telescoped series for $-1 \leqslant x \leqslant +1$.

16. Repeat problem 15 this way: The highest power term in the original Maclaurin series is $3x^3$. In this term *only*, replace the x^3 term by a combination of T_n, truncate the T_3 term, and convert the remaining T_n terms back into powers of x. Add these terms back to the corresponding terms of the original Maclaurin series, truncating the $3x^3$ term. Compare with the result of problem 15. Can this be generalized to other series?

17. At the end of Chapter 1 we showed graphically that if we have a function $f(x)$ which we wish to evaluate at $x = a$, but we happen to use a slightly erroneous value of $\dot{a}$ instead of a, then we make an error

$$\varepsilon_f = f(\dot{a}) - f(a) = \varepsilon_a f'(\dot{a}) = (\dot{a} - a)f'(\dot{a})$$

where ε_f is the error in evaluating $f(x)$ and ε_a is the error made in using $\dot{a}$ instead of a. Prove this result by taking the function $f(x)$, expanding it in a Taylor series about $x_0 = \dot{a}$, and then evaluating the resulting series for $x = a$. If $(a - \dot{a})$ is very small, then the higher terms of the series drop out, and a slight rearrangement gives

$$f(\dot{a}) - f(a) = (\dot{a} - a)f'(\dot{a}).$$

18. Define the quantity $h = \pi/18$ radians $= 0.174532925199433$ (which corresponds to an angle of 10 degrees). Hence we have $\cos h = \cos 10°$ $= 0.984807753012208$. Now use Chebyshev polynomials, and especially their recurrence equation (Eq. 2-11) to compute $\cos nh$ for $n = 2, 3, \cdots,$ 18, and print the results in the form of a cosine table. Compare your results with a good book of tables.

Roots of Equations

We occasionally have an equation of the form

$$f(x) = 0$$

and must find its roots; that is, we must find those values of x for which this equation holds.

The function $f(x)$ may be a simple polynomial such as

$$x^2 - 3x + 2 = 0,$$

which has the two roots $x_1 = 2$ and $x_2 = 1$; for equations as simple as this we do not need a computer at all. On the other hand, we might have the transcendental equation

$$e^x - 3 \cos x = 0,$$

which has an infinite number of roots; finding just one of these by hand is rather tough.

In this chapter we shall show several methods for finding such roots. At the beginning we shall assume no restrictions on the function $f(x)$ except that the roots are real, later we shall discuss the special cases of polynomials and of complex roots.

3.1 Solution by Iteration

In Example 1-3 of Chapter 1, we found the square root of 1.69 by a process called *iteration*. This involved taking an initial guess at the correct answer and, by repeating (*iterating*) a few steps over and over, trying to improve the guess so that we approach the correct answer. We will now see how to use an iterative method to find the roots of equations.

Let us start with the general equation

$$f(x) = 0 \qquad (3\text{-}1)$$

whose root or roots we would like to know. Suppose $\bar{x}$ is one such root. If we insert this value of $\bar{x}$ into $f(x)$, we find that

$$f(\bar{x}) = 0,$$

and the equation is satisfied. If, on the other hand, we try some value of $\dot{x}$ which is *not* a root and insert into $f(x)$, we obtain

$$f(\dot{x}) \neq 0,$$

and the equation is *not* satisfied.

Let us now perform some algebraic manipulations on $f(x)$ so as to extract an x out of it and bring it to the other side of the equation. This will give us some new equation of the form

$$g(x) = x, \qquad (3\text{-}2)$$

where $g(x)$ is a new function, different from the original $f(x)$. As before, if we take a root $\bar{x}$ and substitute into $g(x)$, we obtain

$$g(\bar{x}) = \bar{x}, \qquad (3\text{-}3)$$

and the equation is satisfied. If, on the other hand, we try some other value $\dot{x}$ which is *not* a root, then we obtain

$$g(\dot{x}) \neq \dot{x},$$

and the equation is *not* satisfied. This is only natural, since (3-2) is only a rearranged version of (3-1), not a different equation at all.

Since we have no idea where the real roots of $f(x)$ lie, let us take a guess at a typical root; call this guess x_0. Substituting this value into $g(x)$, we obtain a new number called x_1:

$$x_1 = g(x_0).$$

If, by good luck, we just happened to guess the right value, so that x_0 is indeed a root of $f(x)$, then by (3-3) x_1 would be the same as x_0. But of course this is not too likely. In general x_0 will not be a good guess, and so x_1 will not equal x_0.

If x_1 is different from x_0, then we have two possible cases: either x_1 will be closer to an actual root $\bar{x}$, or else it will be farther away. We have no way of telling yet which it will be, but as a first guess we rather suspect that it will be a matter of chance. This is not quite so, as we shall see.

Let us now take the value of x_1 and again substitute into $g(x)$ to obtain a still different number we shall call x_2:

$$x_2 = g(x_1).$$

x_2 is different from x_1 and might again be either closer to an actual root $\bar{x}$ than x_1, or farther away.

Whether x_1 is closer than x_0 to the actual root or not is actually not quite random, but depends on the derivative of $g(x)$ in a way we shall see in a few pages. Further, whether x_2 is closer than x_1 or not depends on the *same* derivative. Thus, in general, *both* x_1 and x_2 will approach the actual root, or they will *both* be farther away.

We now continue repeating the same process over and over, giving us the following iteration:

$$
\begin{aligned}
x_1 &= g(x_0) \\
x_2 &= g(x_1) \\
x_3 &= g(x_2) \\
x_4 &= g(x_3) \\
&\vdots \\
x_i &= g(x_{i-1}) \\
x_{i+1} &= g(x_i) \\
&\vdots
\end{aligned}
\tag{3-4}
$$

Although there are exceptions, we usually find that the approximate values $x_1, x_2, x_3, x_4, \ldots, x_i, x_{i+1}, \ldots$ either approach toward the actual root $\bar{x}$ so that each x_i is closer than the previous x_{i-1}, or else that these values move away from the actual root so that each is farther away than the previous value.

Generally, though, we have no idea of what the actual root is. How can we tell whether the x_i are *converging to* the root, or *diverging from* the root? We do this indirectly.

In order for this entire process to converge to a root $\bar{x}$, after enough repetitions we must have x_i as close to $\bar{x}$ as desired. That is, suppose we choose an arbitrarily small number ε, as small as we wish. For the iteration to converge to the root $\bar{x}$ it is necessary that x_i eventually approach so close to $\bar{x}$ that the difference $|x_i - \bar{x}|$ is less than our arbitrary ε, and that it remain less for all succeeding x_i. Suppose we need M iterations before this requirement is satisfied. Then we write

$$|x_i - \bar{x}| < \varepsilon \qquad \text{for all } i \geqslant M.$$

Put another way, we may say that, as i approaches infinity, the quantity $(x_i - \bar{x})$ must approach zero:

$$\lim_{i \to \infty} (x_i - \bar{x}) = 0, \qquad (3\text{-}5)$$

or, by adding $\bar{x}$ to both sides, we have

$$\lim_{i \to \infty} x_i = \bar{x}, \qquad (3\text{-}6)$$

which simply means that, as i becomes large, the x_i approach $\bar{x}$ as a limiting value.

Now, if expression (3-5) holds for x_i it will also hold for x_{i+1}, so that we can write

$$\lim_{i \to \infty} (x_{i+1} - \bar{x}) = 0.$$

If expression (3-6) is true, then we can substitute x_i instead of $\bar{x}$ into the limit above:

$$\lim_{i \to \infty} (x_{i+1} - x_i) = 0. \qquad (3\text{-}7)$$

The result can be stated like this: As i becomes very large and approaches ∞, the difference between any two successive approximations x_i and x_{i+1} must approach zero.

This condition is absolutely necessary if our iterative process is to converge to a root $\bar{x}$. It is not necessary if the process diverges, although we may sometimes encounter it during iteration, either because the process is converging to another root or perhaps even a false value, or else because the process has diverged and x_i is the maximum number that can be handled by the computer. A computer which has no indication for overflow will then make all succeeding x_i equally large.

Hence we now have a way of telling whether the iteration is converging or not. We simply write the program in such a way that a maximum number of iterations is specified. If, by the time this number of iterations is done, the difference between two successive x_i is not very small, then the process is apparently diverging. If at any time the difference becomes smaller than some arbitrary small number, even though perhaps the iteration has only been done a few times, then we assume that the answer has been found. Note that this criterion may (very rarely) give us the wrong answer. In any case, whatever answer we receive should be checked by substituting into the original $f(x)$ to see whether $f(x) = 0$.

Before we go on to examine under what conditions the iteration will converge, let us do a short example.

The roots of

$$f(x) = x^2 - 5x + 4 = 0$$

are $x = 1$ and $x = 4$. There are many ways to separate this equation into the form

$$x = g(x).$$

One easy way is to write

$$0 = x^2 - 5x + 4$$

and then move the $5x$ term to the other side:

$$5x = x^2 + 4$$

$$x = \frac{x^2 + 4}{5},$$

which, for the purposes of iteration, becomes

$$x_{i+1} = \frac{(x_i)^2 + 4}{5}.$$

If we take an initial guess $x_0 = 2$, then we have

$$x_1 = \frac{2^2 + 4}{5} = \frac{8}{5} = 1.6,$$

$$x_2 = \frac{(1.6)^2 + 4}{5} = \frac{6.56}{5} = 1.312,$$

$$x_3 = \frac{(1.312)^2 + 4}{5} = \frac{5.721}{5} = 1.144,$$

$$x_4 = \frac{(1.144)^2 + 4}{5} = \frac{5.309}{5} = 1.062,$$

$$x_5 = \frac{(1.062)^2 + 4}{5} = \frac{5.126}{5} = 1.025,$$

$$x_6 = \frac{(1.025)^2 + 4}{5} = \frac{5.051}{5} = 1.010,$$

$$x_7 = \frac{(1.010)^2 + 4}{5} = \frac{5.020}{5} = 1.004.$$

Starting with the initial guess $x_0 = 2$ the procedure is obviously converging to the correct root $x = 1$.

To find the second root $x = 4$ we might try, as a first guess, $x_0 = 5$:

$$x_1 = \frac{5^2 + 4}{5} = \frac{29}{5} = 5.8,$$

$$x_2 = \frac{(5.8)^2 + 4}{5} = \frac{37.64}{5} = 7.528,$$

$$x_3 = \frac{(7.528)^2 + 4}{5} = \frac{60.6}{5} = 12.12.$$

Instead of converging to the correct root of $x = 4$, we see that this time the procedure diverges.

Obviously there are many ways of breaking down an equation

$$f(x) = 0$$

into the form needed for iteration

$$x = g(x),$$

and it is quite possible that some other form might have worked better. We therefore need some way of determining whether or not a particular expression $g(x)$ will converge to a solution or not.

3.2 Convergence Criterion for the Iteration
x = g(x)

To analyze the convergence of an iteration, we shall need the following theorem:

Mean Value Theorem *Suppose a function g(x) and its derivative g'(x) are both continuous in the interval a ⩽ x ⩽ b. Then there exists at least one ξ, a < ξ < b such that*

$$g'(\xi) = \frac{g(b) - g(a)}{b - a}.$$

The proof can be found in any calculus text, and need not concern us here.

Figure 3-1 shows the meaning of the theorem: if a function $g(x)$ is defined in the region between $x = a$ and $x = b$ and is both differentiable (smooth)

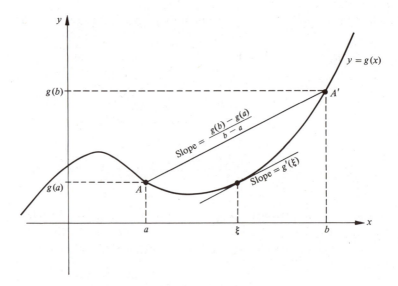

Figure 3-1.   Meaning of the mean value theorem.

and continuous, we may draw a secant line AA' between the points A and A', as shown. The slope of this secant line is

$$\frac{g(b) - g(a)}{b - a}.$$

The mean value theorem then states that there is at least one point on the curve, at $x = \xi$, where the tangent to the curve has the same slope as the line AA'. Physically this is obvious from Figure 3-1, since if the curve deviates from the secant line, then there must be a part of the curve which has a greater slope, and another part which has a smaller slope. If the curve is continuous and smooth (meaning that $g'(x)$ is continuous), then between these two parts must be a point where the curve has neither a greater slope nor a smaller slope, but the slope is exactly equal to the slope of the secant.

For our purposes we shall rewrite the equation of the mean value theorem as

$$g(b) - g(a) = g'(\xi)(b - a). \tag{3-8}$$

We now apply the theorem to the sequence of successive guesses in an iteration,

$$\begin{aligned} x_1 &= g(x_0) \\ x_2 &= g(x_1) \\ x_3 &= g(x_2) \\ &\vdots \\ x_i &= g(x_{i-1}) \\ x_{i+1} &= g(x_i). \end{aligned} \tag{3-4}$$

Now let $a = x_{i-1}$ and let $b = x_i$, and insert into Eq. (3-8):

$$g(x_i) - g(x_{i-1}) = g'(\xi_i)(x_i - x_{i-1}),$$

where ξ_i is in the region between x_i and x_{i-1}.

Since $g(x_i) = x_{i+1}$ and $g(x_{i-1}) = x_i$, we can write the following equation (using absolute values)

$$|x_{i+1} - x_i| = |g'(\xi_i)|\,|x_i - x_{i-1}|.$$

As long as the appropriate ξ_i is used, the mean value theorem can be applied to each iteration of the sequence. Thus we get a set of equations

$$\begin{aligned} |x_2 - x_1| &= |g'(\xi_1)|\,|x_1 - x_0| \\ |x_3 - x_2| &= |g'(\xi_2)|\,|x_2 - x_1| \\ |x_4 - x_3| &= |g'(\xi_3)|\,|x_3 - x_2| \\ &\vdots \\ |x_i - x_{i-1}| &= |g'(\xi_{i-1})|\,|x_{i-1} - x_{i-2}| \\ |x_{i+1} - x_i| &= |g'(\xi_i)|\,|x_i - x_{i-1}|. \end{aligned}$$

In each of these expressions, the $\xi_1, \xi_2, \xi_3, \ldots, \xi_{i-1}, \xi_i$ are different. But suppose that in the entire region which includes the actual root $\bar{x}$ as well as all of the x_i, the derivative $g'(x)$ is bounded. We may then write

$$|g'(\xi_i)| \leqslant M$$

for some upper bound M, and for any i. We may then write

$$|x_2 - x_1| \leqslant M|x_1 - x_0|$$
$$|x_3 - x_2| \leqslant M|x_2 - x_1|$$
$$|x_4 - x_3| \leqslant M|x_3 - x_2|$$
$$\vdots$$
$$|x_i - x_{i-1}| \leqslant M|x_{i-1} - x_{i-2}|$$
$$|x_{i+1} - x_i| \leqslant M|x_i - x_{i-1}|.$$

We now have an entire set of inequalities. Let us take the first and substitute into the second:

$$|x_3 - x_2| \leqslant M|x_2 - x_1| \leqslant M \cdot M|x_1 - x_0|$$
$$|x_3 - x_2| \leqslant M^2|x_1 - x_0|.$$

Substitute this result into the third inequality:

$$|x_4 - x_3| \leqslant M|x_3 - x_2| \leqslant M \cdot M^2|x_1 - x_0|$$
$$|x_4 - x_3| \leqslant M^3|x_1 - x_0|.$$

Now substitute this result into the next, and proceed all the way until the last inequality, so that the final result is

$$|x_{i+1} - x_i| \leqslant M^i|x_1 - x_0|. \tag{3-9}$$

We have already shown, in Eq. (3-7), that for large i the difference between x_{i+1} and x_i must approach zero. Thus the left side of the inequality (3-9) must become very small. This may happen for a number of different reasons having nothing to do with Eq. (3-9) at all. On the other hand, we can force the left side to zero if we make the right side very small. One way to do this is to make $M < 1$, so that M^i approaches zero for large i.

In other words, the iteration *may* converge even if M is large. But if $M < 1$, then the iteration *must* converge. We say that the condition that $M < 1$ is *sufficient* to force convergence, but not *necessary* for convergence.

Hence it is sufficient to show that

$$|g'(x)| \leqslant M < 1 \tag{3-10}$$

for all x in the region including all the intermediate values x_i and the correct value $\bar{x}$. Moreover, if $g'(x)$ is near zero in the entire region the iteration converges quickly; if the derivative is near 1 in magnitude, the iteration converges quite slowly.

Let us return to our previous example

$$x^2 - 5x + 4 = 0$$

$$x = g(x) = \frac{x^2 + 4}{5}$$

$$g'(x) = \frac{2x}{5}.$$

Using an initial value $x_0 = 2$, we find that the derivative is 4/5, and so the convergence starts rather slowly. The derivative is less than 1 in magnitude for all x such that $-5/2 < x < 5/2$, and so the root at $x = 1$ is easily found.

The root $x = 4$, and the initial guess $x_0 = 5$, are both outside the region where the derivative is less than 1 in magnitude, and so the process need not converge; in this case it does not.

There are a few pathological examples where the iteration converges despite the fact that the derivative is greater than 1, as we can see from the following:

$$x^2 - 5x + 4 = 0.$$

Add x to both sides:

$$x = x^2 - 4x + 4 = g(x)$$
$$g'(x) = 2x - 4.$$

The initial value $x_0 = 0$ leads to a derivative of -4, while the derivative at the solution $x = 4$ is $+4$. Yet, if we choose $x_0 = 0$, we obtain

$$x_1 = x_0^2 - 4x_0 + 4 = 0 - 0 + 4 = 4.$$

Obviously we cannot expect such streaks of fortune for long, and so we generally consider the condition that $|g'(x)| < 1$ over the entire region as both sufficient and, generally, necessary.

The requirement that the absolute value of $g'(x)$ need be less than 1 in the region containing the correct root $\bar{x}$ and the values x_i is easily shown from geometric considerations. The iterating equation

$$x_{i+1} = g(x_i)$$

is obviously satisfied at the root $\bar{x}$, so that

$$\bar{x} = g(\bar{x}).$$

This expression can be graphed as the straight line $y = x$, and the curve $y = g(x)$; the root $\bar{x}$ is then that value of x at which the two curves cross, as shown in Figure 3-2.

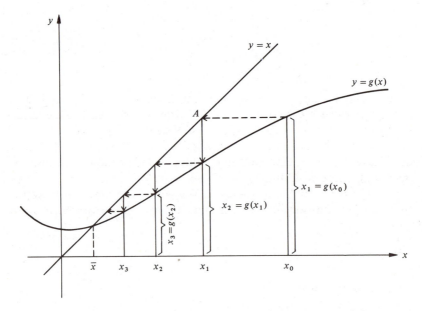

Figure 3-2.   Graphical meaning of the iteration $x_{i+1} = g(x_i)$.

As shown in Figure 3-2, an initial value x_0 is chosen as the first guess, and x_1 is obtained from the curve $y = g(x)$. The value of x_1 is then "mirrored" from the curve $y = x$ at point A, and a new value x_2 is obtained from the curve $y = g(x)$. The case shown in Figure 3-2 shows convergence toward the value $\bar{x}$.

There are four possible cases, as shown in Figure 3-3, of monotonic and oscillating convergence and divergence.

Case a, shown in Figure 3-3a, shows what happens if $g'(x)$ is between 0 and 1; even if x_0 is far from the actual root (which lies at the crossing of the $y = x$ and $y = g(x)$ curves) the successive values of x_i approach the root from one side—this is meant by *monotonic* convergence.

Case b, shown in part b of Figure 3-3, shows the situation when $g'(x)$ is between -1 and 0. Even if x_0 is rather far from the actual root, successive values of x_i approach the root, although they oscillate around it.

Case c, shown in part *c* of Figure 3-3, shows the divergence when $g'(x)$ is greater than $+1$. Successive x_i all move away from the root monotonically.

Case d, shown in part *d* of Figure 3-3, shows oscillating divergence when $g'(x) < -1$, so that $|g'(x)| > 1$.

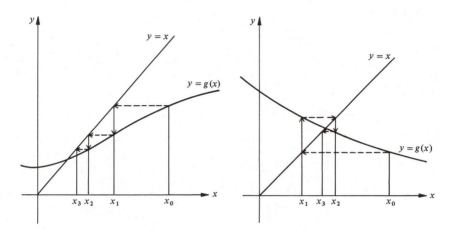

a. Monotonic convergence   b. Oscillating convergence

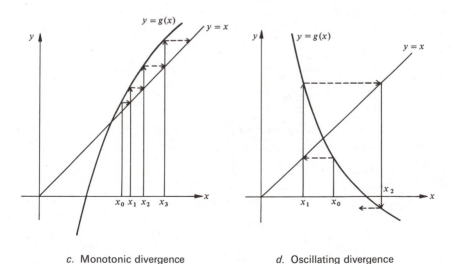

c. Monotonic divergence   d. Oscillating divergence

Figure 3-3. Four possible cases of convergence and divergence in the iteration $x = g(x)$.

3.3 Example 3-1. Simple Iteration

As an example of iterative finding of the roots of an equation, not necessarily a polynomial equation, consider the problem shown in Figure 3-4. Two curves, $y_1 = e^x$ and $y_2 = 3x$ intersect in the two places shown. The problem is to find the x coordinates of the two intersections.

At the intersections

$$y_1 = y_2,$$
$$e^x = 3x,$$
$$f(x) = e^x - 3x = 0,$$

and so we are seeking the two roots of $f(x) = 0$.

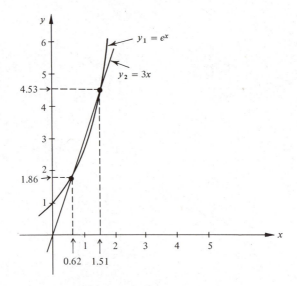

Figure 3-4.   Graph for Example 3-1.

To separate into the form $x = g(x)$, we write

$$0 = e^x - 3x$$
$$3x = e^x$$
$$x = \frac{e^x}{3},$$

or

$$x_{i+1} = \frac{e^{x_i}}{3}.$$

Note that this is only one of the many ways of writing $x = g(x)$, and not necessarily the best—only the most obvious.

To find the root at $x = 0.62$, we might try an initial guess $x_0 = 0$ or perhaps $x_0 = 1$. Using the value $x_0 = 0$, we have

$$x_1 = \frac{e^0}{3} = 0.333\ldots,$$

$$x_2 = \frac{e^{0.333}}{3} = 0.465,$$

$$x_3 = \frac{e^{0.465}}{3} = 0.530,$$

$$x_4 = \frac{e^{0.530}}{3} = 0.567,$$

which seems to converge quite well.

To get the root at $x = 1.51$, we might try an initial guess $x_0 = 2$:

$$x_1 = \frac{e^2}{3} = 2.46,$$

$$x_2 = \frac{e^{2.46}}{3} = 3.91,$$

$$x_3 = \frac{e^{3.91}}{3} = 16.7,$$

which diverges very quickly since x_4 would be almost 6 million.

The reason for the divergence is quite obvious if we examine the derivative $g'(x)$.

$$g(x) = \frac{e^x}{3}$$

$$g'(x) = \frac{e^x}{3}.$$

To find the range of x for which the absolute value of the derivative is less than 1, we let

$$\left| \frac{e^x}{3} \right| < 1$$

$$e^x < 3$$

$$x < \ln 3 \approx 1.1.$$

Whenever $x < 1.1$ the process will converge. Thus the initial guess $x_0 = 2$ will probably cause divergence, but moreover the solution $x = 1.51$ itself is out of the range of guaranteed convergence, and so most likely cannot be found by this method.

The following FORTRAN program demonstrates this fact. Starting with four initial values of x_0,

$$\left.\begin{array}{l} x_a = -1 \\ x_b = 0 \\ x_c = +1 \end{array}\right\} \text{ Convergent}$$

$$x_d = +2 \quad \text{Divergent}$$

the program performs all four iterations at the same time. As expected, starting values below 1.1 all converge to the root at approximately 0.62, while the starting value of $x_0 = +2$ diverges very quickly.

The iteration program follows:

```
      G(X) = EXP (X) / 3.
      XA = -1.
      XB = 0.
      XC = 1.
      XD = 2.
      WRITE (3, 5) XA, XB, XC, XD
    5 FORMAT (4X, 3F15.8, E17.8)
      DO 10 I = 1, 40
      XA = G(XA)
      XB = G(XB)
      XC = G(XC)
      XD = G(XD)
      WRITE (3, 20) I, XA, XB, XC, XD
   20 FORMAT (' X', I2, 3F15.8, E17.8)
   10 CONTINUE
      CALL EXIT
      END
```

As we can see from the following results, starting values of -1, 0, and $+1$ require 38, 37, and over 40 iterations, respectively, to arrive at an answer accurate to eight places; the iteration starting at $+2$ does not converge at all.

	-1.00000000	0.00000000	1.00000000	0.20000000E 01
X 1	0.12262648	0.33333333	0.90609394	0.24630187E 01
X 2	0.37682069	0.46520414	0.82487918	0.39133994E 01
X 3	0.48588097	0.53077973	0.76053503	0.16689622E 02
X 4	0.54186883	0.56675251	0.71314019	0.59032312E 07
X 5	0.57307226	0.58751131	0.68012947	0.56713727E 38
X 6	0.59123599	0.59983481	0.65804443	0.56713727E 38
X 7	0.60207317	0.60727261	0.64367080	0.56713727E 38
X 8	0.60863342	0.61180622	0.63448509	0.56713727E 38
X 9	0.61263934	0.61458621	0.62868358	0.56713727E 38
X10	0.61509844	0.61629712	0.62504683	0.56713727E 38
X11	0.61661290	0.61735246	0.62277781	0.56713727E 38
X12	0.61754744	0.61800432	0.62136632	0.56713727E 38
X13	0.61812483	0.61840730	0.62048989	0.56713727E 38
X14	0.61848183	0.61865656	0.61994631	0.56713727E 38

X15	0.61870267	0.61881079	0.61960941	0.56713727E 38
X16	0.61883932	0.61890623	0.61940070	0.56713727E 38
X17	0.61892389	0.61896530	0.61927144	0.56713727E 38
X18	0.61897623	0.61900187	0.61919139	0.56713727E 38
X19	0.61900863	0.61902450	0.61914183	0.56713727E 38
X20	0.61902869	0.61903851	0.61911115	0.56713727E 38
X21	0.61904111	0.61904719	0.61909215	0.56713727E 38
X22	0.61904879	0.61905256	0.61908039	0.56713727E 38
X23	0.61905355	0.61905588	0.61907311	0.56713727E 38
X24	0.61905650	0.61905794	0.61906861	0.56713727E 38
X25	0.61905832	0.61905921	0.61906582	0.56713727E 38
X26	0.61905945	0.61906000	0.61906409	0.56713727E 38
X27	0.61906015	0.61906049	0.61906302	0.56713727E 38
X28	0.61906058	0.61906079	0.61906236	0.56713727E 38
X29	0.61906085	0.61906098	0.61906195	0.56713727E 38
X30	0.61906101	0.61906109	0.61906169	0.56713727E 38
X31	0.61906112	0.61906117	0.61906154	0.56713727E 38
X32	0.61906118	0.61906121	0.61906144	0.56713727E 38
X33	0.61906122	0.61906124	0.61906138	0.56713727E 38
X34	0.61906124	0.61906125	0.61906134	0.56713727E 38
X35	0.61906126	0.61906127	0.61906132	0.56713727E 38
X36	0.61906127	0.61906127	0.61906131	0.56713727E 38
X37	0.61906127	0.61906128	0.61906130	0.56713727E 38
X38	0.61906128	0.61906128	0.61906129	0.56713727E 38
X39	0.61906128	0.61906128	0.61906129	0.56713727E 38
X40	0.61906128	0.61906128	0.61906129	0.56713727E 38

We now ask whether a faster process might be found that requires fewer iterations for the same accuracy, and whether there is a simple way of judging how fast a given $g(x)$ might converge.

3.4 Order of Convergence

As already stated, the magnitude of $g'(x)$ at the solution $x = \bar{x}$ has a great bearing on whether a process converges, and if so, how fast. If this derivative has a very small value, or is zero, we might expect a very fast convergence. We now show that this is so.

Let ε_i be the error in the ith iteration; that is,

$$\varepsilon_i = x_i - \bar{x}.$$

If we know the function $g(x)$ and its derivatives at the correct root $\bar{x}$ we can expand about $\bar{x}$ in a Taylor series and find the value of the function at x_i:

$$g(x_i) = g(\bar{x}) + g'(\bar{x})(x_i - \bar{x}) + g''(\bar{x})\frac{(x_i - \bar{x})^2}{2!} + g'''(\bar{x})\frac{(x_i - \bar{x})^3}{3!} + \cdots,$$

or

$$g(x_i) - g(\bar{x}) = g'(\bar{x})(x_i - \bar{x}) + g''(\bar{x})\frac{(x_i - \bar{x})^2}{2!} + g'''(\bar{x})\frac{(x_i - \bar{x})^3}{3!} + \cdots.$$

Since

$$x_{i+1} = g(x_i)$$
$$\bar{x} = g(\bar{x})$$
$$\varepsilon_i = x_i - \bar{x},$$

we can also write

$$x_{i+1} - \bar{x} = g'(\bar{x})\varepsilon_i + g''(\bar{x})\frac{\varepsilon_i^2}{2!} + g'''(\bar{x})\frac{\varepsilon_i^3}{3!} + \cdots.$$

The left side is the error in the $(i + 1)$th iteration, and is therefore expressed by ε_{i+1}, so that

$$\varepsilon_{i+1} = g'(\bar{x})\varepsilon_i + g''(\bar{x})\frac{\varepsilon_i^2}{2!} + g'''(\bar{x})\frac{\varepsilon_i^3}{3!} + \cdots.$$

We have now expressed the error in the $(i + 1)$th iteration in terms of the ith error. For fast convergence, we would like each error to be much smaller than the previous error.

As long as the error ε_i is fairly large, there is little we can say about convergence. But when ε_i becomes fairly small, perhaps after the first few iterations, we see that ε_i^2 is much smaller, ε_i^3 is smaller yet, and so on, so that the first non-zero term of this series overshadows the following terms.

If the first derivative were zero, each error would be roughly proportional to the square of the previous error; if the first two derivatives were zero, then each error would be roughly proportional to the cube of the previous error, and convergence would be even faster. We can therefore define the following as a measure of the speed of convergence:

Definition *The order of convergence is the order of the lowest non-zero derivative of $g(x)$ at the solution $\bar{x}$.*

Some simple examples: if

$$f(x) = e^x - 3x = 0,$$

then

$$g(x) = x = \frac{e^x}{3},$$

$$g'(x) = \frac{e^x}{3}.$$

This derivative is never zero, so that the lowest nonzero derivative is the first;

hence this is first-order convergence, also sometimes called *linear* convergence. Likewise, if

$$f(x) = x^2 - 5x + 4 = 0$$

then
$$g(x) = x = \frac{x^2 + 4}{5}$$

$$g'(x) = \frac{2x}{5}.$$

This derivative is nonzero at either of the two solutions $x = 1$ or $x = 4$, and so the order is first order.

3.5 The Newton-Raphson Method

We now show another method of iteration which is second order for simple roots. It is essentially similar to iteration of the type

$$x = g(x)$$

except that in this case the $g(x)$ is chosen so that its first derivative at the root $\bar{x}$ vanishes.

Figure 3-5 shows the original equation $f(x) = 0$; the desired root $\bar{x}$ is the x coordinate where the curve $f(x)$ crosses the x axis.

Suppose that in Figure 3-5 we pick an initial guess x_0. To get a new value x_1, we draw the tangent to the curve at the point $x = x_0$, $y = f(x_0)$, and follow the tangent down to its intersection with the x axis. To get x_2 we repeat the process, starting from x_1.

As seen from Figure 3-5,

$$x_1 = x_0 - \Delta x.$$

The slope of the tangent to the curve at the point $[x_0, f(x_0)]$ is

$$f'(x_0) = \frac{f(x_0)}{\Delta x},$$

so that

$$\Delta x = \frac{f(x_0)}{f'(x_0)},$$

and

$$x_1 = x_0 - \frac{f(x_0)}{f'(x_0)},$$

or in general

$$x_{i+1} = x_i - \frac{f(x_i)}{f'(x_i)} = g(x_i), \qquad (3\text{-}11)$$

which is the formula for a Newton-Raphson iteration.

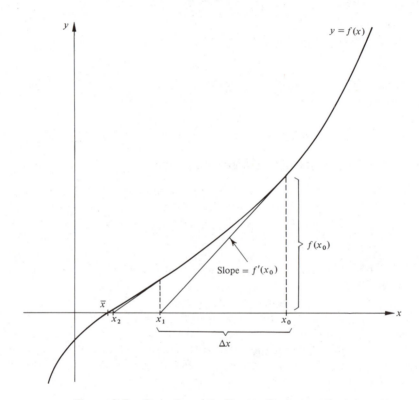

Figure 3-5.   Derivation of the Newton-Raphson method.

For mathematical thoroughness, we may arrive at the same equation through another, less intuitive approach. Suppose $f(x)$ is continuous and differentiable, so that it may be expanded in a Taylor series. We may then express $f(x_{i+1})$ in terms of the function and its derivatives at x_i as follows:

$$f(x_{i+1}) = f(x_i) + f'(x_i)(x_{i+1} - x_i) + f''(x_i)\frac{(x_{i+1} - x_i)^2}{2!} + \cdots.$$

Now suppose that x_{i+1} is very close to the actual root $\bar{x}$ so that $f(x_{i+1})$ is almost equal to zero; also suppose that x_i is also very close to x_{i+1}, so that $(x_{i+1} - x_i)^2$ and all higher powers can be neglected. Then the above Taylor series simplifies to

$$0 = f(x_i) + f'(x_i)(x_{i+1} - x_i),$$

which can be solved for x_{i+1} to give

$$x_{i+1} = x_i - \frac{f(x_i)}{f'(x_i)},$$

as before.

To show that the Newton-Raphson method provides second-order convergence, we would like to show that the first derivative of $g(x) = x - f(x)/f'(x)$ is zero, whereas the second derivative is not zero. We start by differentiating

$$g(x) = x - \frac{f(x)}{f'(x)}$$

$$g'(x) = 1 - \frac{f'(x)f'(x) - f(x)f''(x)}{[f'(x)]^2}.$$

Simplifying, we have

$$g'(x) = \frac{[f'(x)]^2 - [f'(x)]^2 + f(x)f''(x)}{[f'(x)]^2}$$

$$= \frac{f(x)f''(x)}{[f'(x)]^2}.$$

At $x = \bar{x}$, we have

$$g'(\bar{x}) = \frac{f(\bar{x})f''(\bar{x})}{[f'(\bar{x})]^2}. \qquad (3\text{-}12)$$

Since $f(\bar{x}) = 0$, the numerator is zero. Hence $g'(\bar{x})$ is zero *if* the denominator is nonzero.

If the root is a simple root, $f'(\bar{x})$ is nonzero; hence for simple roots the Newton-Raphson process is second order (since we could easily show that the second derivative $g''(\bar{x})$ is not zero).

Consider, however, two roots which are very close together. The mean value theorem then states that between the two roots there must be a value ξ such that $f'(\xi) = 0$. If the two roots are moved closer together, until in the limit they coincide, then the derivative at the root itself becomes zero. Thus both the numerator and the denominator of Eq. (3-12) are zero, and $g'(\bar{x})$ might be zero, finite, or infinite, depending on whether the numerator goes to zero faster than the denominator, whether they go to zero at the same rate, or whether the denominator goes to zero faster.

To analyze this problem, we introduce a method called L'Hopital's rule:

L'Hopital's Rule *Suppose that two functions $F(x)$ and $G(x)$ both approach 0 as x approaches a. That is,*

$$\lim_{x \to a} F(x) = 0$$

$$\lim_{x \to a} G(x) = 0.$$

To find the limiting value of $F(x)/G(x)$ as x approaches a, we find the limit of $F'(x)/G'(x)$:

$$\lim_{x \to a} \frac{F(x)}{G(x)} = \lim_{x \to a} \frac{F'(x)}{G'(x)}.$$

This limit can be either finite or infinite.

Applying the rule to Eq. (3-12), we have

$$g'(\bar{x}) = \lim_{x \to \bar{x}} g'(x) = \lim_{x \to \bar{x}} \frac{f(x)f''(x)}{[f'(x)]^2}$$

$$= \lim_{x \to \bar{x}} \frac{d/dx[f(x)f''(x)]}{d/dx\{[f'(x)]^2\}}$$

$$= \lim_{x \to \bar{x}} \frac{f'(x)f''(x) + f(x)f'''(x)}{2f'(x)f''(x)}.$$

Inserting the limiting value $x = \bar{x}$ and breaking up into two fractions, we have

$$g'(\bar{x}) = \frac{f'(\bar{x})f''(\bar{x})}{2f'(\bar{x})f''(\bar{x})} + \frac{f(\bar{x})f'''(\bar{x})}{2f'(\bar{x})f''(\bar{x})}$$

$$= \frac{1}{2} + \frac{f(\bar{x})f'''(\bar{x})}{2f'(\bar{x})f''(\bar{x})}. \tag{3-13}$$

If $\bar{x}$ is a double root, then $f(\bar{x})$ and $f'(\bar{x})$ are both zero, and so we again have a problem—the fraction on the right is 0/0. To evaluate it, we again apply L'Hopital's rule. Differentiating the numerator and denominator and again inserting $x = \bar{x}$, we obtain

$$\frac{f(\bar{x})f'''(\bar{x})}{2f'(\bar{x})f''(\bar{x})} = \frac{f(\bar{x})f^{IV}(\bar{x}) + f'(\bar{x})f'''(\bar{x})}{2[f'(x)f'''(\bar{x}) + f''(\bar{x})f''(\bar{x})]} = 0.$$

The latter expression is zero since $f(\bar{x})$ and $f'(\bar{x})$ are zero, so that the numerator is zero, while the $f''(\bar{x}) f''(\bar{x})$ in the denominator is nonzero for a double root. Hence the expression for $g'(\bar{x})$ in Eq. (3-13) is equal to $\frac{1}{2}$, and the Newton-Raphson method is only first order for double roots.

For triple roots and higher-order roots, we would find that $f''(\bar{x})$ is also zero, and so we would have to perform further differentiations in an effort to apply L'Hopital's rule again; eventually, however, we would see that $g'(\bar{x})$ would again be nonzero, and the process is still first order.

To summarize the results of this involved argument: the Newton-Raphson method is a second-order iteration for simple first-order roots, and is only first-order for multiple roots. Hence it will take longer to find a multiple root than a simple root.

3.6 Example 3-2. Newton-Raphson Iteration

We now use the Newton-Raphson method to get the solution of the equation from Example 3-1:

$$f(x) = e^x - 3x = 0,$$

where $f'(x) = e^x - 3$, and therefore the iteration formula is

$$x_{i+1} = x_i - \frac{f(x_i)}{f'(x_i)} = x_i - \frac{e^{x_i} - 3x_i}{e^{x_i} - 3}.$$

As in Example 3-1, the following program starts with the four initial values of

$$x_a = -1,$$
$$x_b = 0,$$
$$x_c = +1,$$
$$x_d = +2.$$

and iterates all four simultaneously:

```
G(X) = X - (EXP (X) - 3. * X) / (EXP (X) - 3.)
XA = -1.
XB = 0.
XC = 1.
XD = 2.
WRITE (3, 5) XA, XB, XC, XD
5 FORMAT (4X, 4F15.8)
DO 10 I = 1, 10
XA = G(XA)
XB = G(XB)
XC = G(XC)
XD = G(XD)
WRITE (3, 20) I, XA, XB, XC, XD
20 FORMAT (' X', I2, 4F15.8)
10 CONTINUE
CALL EXIT
END
```

The following results are produced:

	-1.00000000	0.00000000	1.00000000	2.00000000
X 1	0.27953084	0.50000000	0.00000000	1.68351826
X 2	0.56800734	0.61005965	0.50000000	1.54348197
X 3	0.61716729	0.61899677	0.61005965	1.51348862
X 4	0.61905838	0.61906128	0.61899677	1.51213725
X 5	0.61906128	0.61906128	0.61906128	1.51213455
X 6	0.61906128	0.61906128	0.61906128	1.51213455
X 7	0.61906128	0.61906128	0.61906128	1.51213455
X 8	0.61906128	0.61906128	0.61906128	1.51213455
X 9	0.61906128	0.61906128	0.61906128	1.51213455
X10	0.61906128	0.61906128	0.61906128	1.51213455

As we can see, only 4 or 5 iterations are required to produce acceptable answers accurate to 8 decimal places, instead of the 38 or 39 iterations required in Example 3-1. The starting values of -1, 0, and $+1$ produce the first root, while the starting value of $+2$ produces the second root, which was not produced by the program of Example 3-1.

3.7 Pitfalls of the Newton-Raphson Method

When the Newton-Raphson method works, it produces fast results in relatively few iterations. Because the convergence for simple roots is second order and each error ε_i is proportional to the square of the previous error ε_{i-1} because of the order, for small errors we will approximately double the number of significant digits on each iteration.

To show this, assume that the error on one iteration is 10^{-n}, represented by $0.0000\ldots0001$, with $n-1$ zeroes between the decimal point and the 1. The next error, which is proportional to the square of the previous error, is therefore approximately 10^{-2n}, which has $2n-1$ zeroes, the next error has $4n-1$ zeroes, and so on. Hence each iteration approximately doubles the number of correct digits.

Sometimes, however, the Newton-Raphson method does not converge, but instead oscillates back and forth. As shown in Figure 3-6, this happens if there is no real root, as in a, if, as in b, there is a symmetry in $f(x)$ around the point $\bar{x}$, or if the initial guess x_0 is so far away from the correct root that some other part of the function "catches" the iteration, as in c.

This raises the problem of when to stop an iteration. There are three criteria:

1. When $|x_{i+1} - x_i| \leqslant m$, where m is some error term which indicates the accuracy we would like. This is generally a valid test, since the difference between two successive iterations must go to zero if the process is to converge. If, on the other hand, the convergence is very slow or else roundoff error is very severe, it is possible that the difference between two successive terms is very small even though we are nowhere near the right answer.

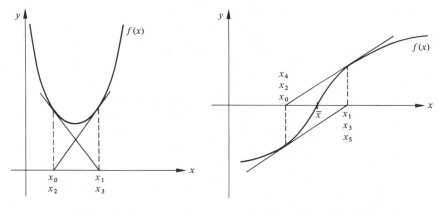

a. No real root.

b. Second derivative $f''(\bar{x}) = 0$.

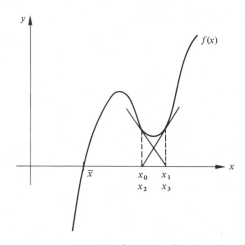

c. x_0 not close enough.

Figure 3-6.   Three functions $f(x)$ which would not lead to convergence with the Newton-Raphson method.

2. In addition to the above test, we should require that $f(x_i) = 0$, or at least, that $|f(x_i)| < \varepsilon$ for some arbitrarily chosen ε. Since $f(x_i)$ is available as part of the calculations for the Newton-Raphson method, this is an easy test.

3. In case neither of the two above tests stops the iteration after some predetermined number of loops, we may assume that the iteration is not converging, and should therefore provide an error exit which prints an error message. This is easily done by placing the entire process into a DO loop.

3.8 The Method of False Position

The method of *false position*, also sometimes called *regula falsi* through a self-perpetuating habit, has certain advantages over the Newton-Raphson method in that it always converges, but not as quickly. Furthermore, it requires two initial guesses instead of just one.

Figure 3-7 shows the method. Suppose that two initial guesses are available, such that x_L is known to be to the left of the actual root $\bar{x}$, and x_R is known to be to the right of the root.

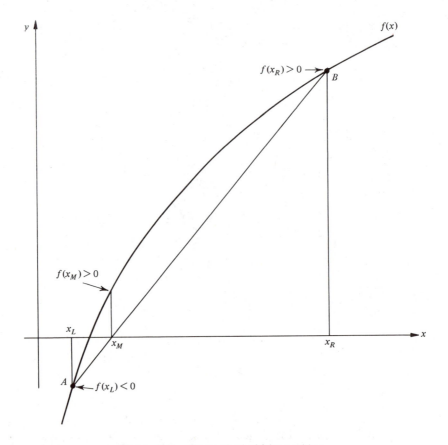

Figure 3-7.   The method of false position.

A line is now drawn from point A, which has the coordinates $(x_L, f(x_L))$, to point B, which has the coordinates $(x_R, f(x_R))$, and the intersection of this line with the x axis is called x_M.

The value of $f(x_M)$ is now examined, and compared with the values of the function at x_L and x_R. Since x_L and x_R are on opposite sides of the root $\bar{x}$, the signs of $f(x_L)$ and $f(x_R)$ are different, indicating that points A and B are on opposite sides of the x axis.

We now compare the sign of $f(x_M)$ with $f(x_L)$. If the signs are the same, then the root cannot be between them; if the signs are different, then the root lies between x_L and x_M. In this case, x_M becomes the new x_R. We now draw a new secant line between x_L and the new x_R and repeat the process. In this way we narrow down the interval in which the root may lie.

The rule is therefore as follows:

1. If $f(x_L) \cdot f(x_M) < 0$, replace x_R with x_M.
2. If $f(x_R) \cdot f(x_M) > 0$, replace x_L with x_M.

The expression for x_M in terms of x_L and x_R is

$$x_M = x_L - \frac{x_R - x_L}{f(x_R) - f(x_L)} f(x_L) = \frac{x_L f(x_R) - x_R f(x_L)}{f(x_R) - f(x_L)}. \quad (3\text{-}14)$$

The method of false position is fairly reliable, but not too fast, though the calculation of x_M becomes tricky when x_L and x_R are close together. Efforts are sometimes made to speed up convergence in monotonically convergent cases by altering terms and trying to force x_M closer to the actual root, but these efforts are often harder than merely repeating a few more iterations. (See problem 11.)

3.9 Example 3-3. Method of False Position

We now try a program to solve

$$f(x) = e^x - 3x = 0$$

by the method of false position. To find the root at 0.62, we start with initial values $x_L = 0$ and $x_R = 1$ with the following program:

```
      F(X) = EXP (X) - 3. * X
      WRITE (3, 5)
    5 FORMAT ('    NO.         XL              XM              XR')
      XL = 0.
      XR = 1.
      DO 100 I = 1, 1000
      XM = (XL * F(XR) - XR * F(XL)) / (F(XR) - F(XL))
      WRITE (3, 10) I, XL, XM, XR
   10 FORMAT (I6, 3F15.8)
      IF (F(XM) * F(XL)) 20, 30, 40
   20 XR = XM
      GO TO 50
   30 XR = XM
   40 XL = XM
```

```
 50 IF (XL - XR) 100, 200, 100
100 CONTINUE
    WRITE (3, 150)
150 FORMAT (' NO ANSWER FOUND AFTER 1000 REPETITIONS.')
    CALL EXIT
200 WRITE (3, 210) XM
210 FORMAT (' THE ANSWER IS ', F15.8)
    CALL EXIT
    END
```

The printout shows the locations of x_L, x_M, and x_R at each iteration. At the 17th iteration x_M reaches the root, and the product F(XM) $*$ F(XL) becomes zero. The IF statement in the program sends control to statement number 30, and x_L, x_M, and x_R become equal; as a result statement number 50 bypasses the DO loop and sends the program to step 200.

Note that the convergence here is about halfway between first-order convergence (38 iterations in Example 3-1) and second-order convergence (about 5 iterations in Example 3-2.)

NO.	XL	XM	XR
1	0.00000000	0.78020271	1.00000000
2	0.00000000	0.67334686	0.78020271
3	0.00000000	0.63568161	0.67334686
4	0.00000000	0.62399050	0.63568161
5	0.00000000	0.62050908	0.62399050
6	0.00000000	0.61948530	0.62050908
7	0.00000000	0.61918536	0.61948530
8	0.00000000	0.61909758	0.61918536
9	0.00000000	0.61907190	0.61909758
10	0.00000000	0.61906439	0.61907190
11	0.00000000	0.61906219	0.61906439
12	0.00000000	0.61906155	0.61906219
13	0.00000000	0.61906136	0.61906155
14	0.00000000	0.61906130	0.61906136
15	0.00000000	0.61906129	0.61906130
16	0.00000000	0.61906128	0.61906129
17	0.00000000	0.61906128	0.61906128

THE ANSWER IS   0.61906128

If we next restart the same program with $x_L = 1$ and $x_R = 2$, we find the second root after 28 iterations.

NO.	XL	XM	XR
1	1.00000000	1.16861533	2.00000000
2	1.16861533	1.31151655	2.00000000
3	1.31151655	1.40666467	2.00000000
4	1.40666467	1.46017025	2.00000000
5	1.46017025	1.48741059	2.00000000
6	1.48741059	1.50057378	2.00000000
7	1.50057378	1.50677349	2.00000000
8	1.50677349	1.50965812	2.00000000
9	1.50965812	1.51099268	2.00000000
10	1.51099268	1.51160848	2.00000000
11	1.51160848	1.51189228	2.00000000
12	1.51189228	1.51202299	2.00000000
13	1.51202299	1.51208319	2.00000000
14	1.51208319	1.51211090	2.00000000
15	1.51211090	1.51212366	2.00000000

16	1.51212366	1.51212954	2.00000000
17	1.51212954	1.51213224	2.00000000
18	1.51213224	1.51213349	2.00000000
19	1.51213349	1.51213406	2.00000000
20	1.51213406	1.51213432	2.00000000
21	1.51213432	1.51213445	2.00000000
22	1.51213445	1.51213450	2.00000000
23	1.51213450	1.51213453	2.00000000
24	1.51213453	1.51213454	2.00000000
25	1.51213454	1.51213454	2.00000000
26	1.51213454	1.51213455	2.00000000
27	1.51213455	1.51213455	2.00000000
28	1.51213455	1.51213455	1.51213455

```
THE ANSWER IS    1.51213455
```

3.10 The Bisection Method

The bisection method is very similar to the method of false position, but somewhat simpler and more foolproof.

As in the method of false position, two initial guesses x_L and x_R are required. The value of x_M, however, is not calculated by drawing a line between the functions at x_L and x_R; instead, x_M is taken exactly halfway between x_L and x_R.

Depending on the exact function, the bisection method may be slightly faster or slightly slower than the method of false position. Its one great advantage is that it provides a definite knowledge of the accuracy (neglecting roundoff). That is, after every iteration the size of the starting interval is reduced by a factor of 2; after n iterations the interval is reduced by a factor of 2^n. Hence if our starting interval (distance between x_L and x_R) is 4, and we need accuracy to within 0.000001, then 22 iterations will reduce the interval by a factor of 2^{22}, or better than 4 million. Hence we know ahead of time just how long the iteration will take and how many iterations are required. No elaborate error testing is therefore necessary, and a simple DO loop can be used to count off the iterations.

The only problems with both the methods of false position and bisection are:

1. The need for two initial guesses, one on each side of the root. This becomes extremely difficult to provide if we either have no idea at all of the root location, or if there are two nearly equal roots. In the latter case, one of our original guesses must be in the space between the two roots, or else we shall miss both roots entirely. Moreover, even order multiple roots are impossible altogether.
2. Roundoff and quantizing errors can cause problems as x_M approaches the actual root $\bar{x}$ when, because of roundoff or quantizing error, $f(x_M)$ may be slightly in error. Since we normally only look at the sign, this does not matter until x_M is so close to the root that it is positive when it should be

negative or vice versa, or accidentally happens to be zero. Generally this happens only when we are very close to the actual answer.

3. When x_M, x_L, and x_R are very close to the actual root $\bar{x}$, $f(x_M)$, $f(x_L)$, and $f(x_R)$ become very small. The common way of testing for different signs is by multiplication of the two numbers to be tested; if the product is negative, then the two numbers had different signs. If, however, the two numbers are both very small, an underflow may cause a zero product, and this test will fail.

Despite all these drawbacks, the method of bisection has the great advantages of requiring little thought, and of providing guaranteed convergence (if two original guesses are available) and almost guaranteed accuracy.

3.11 Example 3-4. Bisection Method

We again solve

$$f(x) = e^x - 3x = 0$$

this time by the method of bisection, with the following program:

```
      F(X) = EXP (X) - 3. * X
      WRITE (3, 5)
    5 FORMAT ('     NO.          XL          XM          XR')
      XL = 0.
      XR = 1.
      DO 100 I = 1, 30
      XM = (XL + XR) / 2.
      WRITE (3, 10) I, XL, XM, XR
   10 FORMAT (I6, 3F15.8)
      IF (F(XM) * F(XL)) 20, 30, 40
   20 XR = XM
      GO TO 50
   30 XR = XM
   40 XL = XM
   50 IF (XL - XR) 100, 200, 100
  100 CONTINUE
  200 WRITE (3, 210) XM
  210 FORMAT (' THE ANSWER IS ', F15.8)
      CALL EXIT
      END
```

Starting with initial guesses $x_L = 0$ and $x_R = 1$, we find the root at $x = 0.62$ after 30 iterations; compare this with the 17 iterations required by the method of false position (Example 3-3).

NO.	XL	XM	XR
1	0.00000000	0.50000000	1.00000000
2	0.50000000	0.75000000	1.00000000
3	0.50000000	0.62500000	0.75000000
4	0.50000000	0.56250000	0.62500000
5	0.56250000	0.59375000	0.62500000

6	0.59375000	0.60937500	0.62500000
7	0.60937500	0.61718750	0.62500000
8	0.61718750	0.62109375	0.62500000
9	0.61718750	0.61914062	0.62109375
10	0.61718750	0.61816406	0.61914062
11	0.61816406	0.61865234	0.61914062
12	0.61865234	0.61889648	0.61914062
13	0.61889648	0.61901855	0.61914062
14	0.61901855	0.61907959	0.61914062
15	0.61901855	0.61904907	0.61907959
16	0.61904907	0.61906433	0.61907959
17	0.61904907	0.61905670	0.61906433
18	0.61905670	0.61906051	0.61906433
19	0.61906051	0.61906242	0.61906433
20	0.61906051	0.61906147	0.61906242
21	0.61906051	0.61906099	0.61906147
22	0.61906099	0.61906123	0.61906147
23	0.61906123	0.61906135	0.61906147
24	0.61906123	0.61906129	0.61906135
25	0.61906123	0.61906126	0.61906129
26	0.61906126	0.61906127	0.61906129
27	0.61906127	0.61906128	0.61906129
28	0.61906128	0.61906128	0.61906129
29	0.61906128	0.61906128	0.61906128
30	0.61906128	0.61906128	0.61906128

THE ANSWER IS   0.61906128

Starting with the initial guesses $x_L = 1$ and $x_R = 2$, we find the root at $x = 1.51$ after only 18 iterations. In this particular case the bisection method gives faster convergence than the method of false position (Example 3-3).

NO.	XL	XM	XR
1	1.00000000	1.50000000	2.00000000
2	1.50000000	1.75000000	2.00000000
3	1.50000000	1.62500000	1.75000000
4	1.50000000	1.56250000	1.62500000
5	1.50000000	1.53125000	1.56250000
6	1.50000000	1.51562500	1.53125000
7	1.50000000	1.50781250	1.51562500
8	1.50781250	1.51171875	1.51562500
9	1.51171875	1.51367187	1.51562500
10	1.51171875	1.51269531	1.51367187
11	1.51171875	1.51220703	1.51269531
12	1.51171875	1.51196289	1.51220703
13	1.51196289	1.51208496	1.51220703
14	1.51208496	1.51214599	1.51220703
15	1.51208496	1.51211547	1.51214599
16	1.51211547	1.51213073	1.51214599
17	1.51213073	1.51213836	1.51214599
18	1.51213073	1.51213455	1.51213836

THE ANSWER IS   1.51213455

3.12 Finding Initial Values

When using any of the previous methods for finding roots of the equation $f(x) = 0$, we need one or two initial guesses; for the bisection methods and the method of false position, the two initial guesses must moreover be on opposite sides of the root $\bar{x}$ to be found. This may raise some problems.

Generally, the equation whose root we need has some physical significance, and so we can guess, from physical reasoning, at an approximate value. If this guess is too far off, however, we have seen that the Newton-Raphson method may not converge, and so even here we would like a more assuring method.

Unfortunately there is no easy way. The initial guesses must often be obtained by a brute force searching method. That is, a program is written to start with an initial guess x_g which is at the very edge of the region which (we think) contains the root, and search through the entire region in steps of size h using the expression

$$x_n = x_g + nh, \qquad n = 0, 1, 2, 3, \ldots, m,$$

with m chosen so that $x_g + mh$ brings us to the opposite edge of the region which (we still hope) has the root.

At each x_n we calculate the value of the function $f(x_n)$ and compare with the value at x_{n-1}; if the two values have opposite signs, then we have just crossed over a root.

There are two difficulties with this procedure. One of them entails the painful choice of m and h. To make sure we do not miss a root, we should use many steps (large m) of small length (small h), but this is apt to take a long time and be expensive. Large steps, on the other hand, are fast but may miss roots if there is an even number of them (as in Figure 3-8a) very close together.

Closely spaced roots can generally be found by taking many tiny search steps. But even more important is the finding of multiple roots, which may be missed by a simple search routine altogether. Figure 3-8b shows two possible cases of multiple roots.

Since in Figure 3-8b $f(x_4)$ is positive and $f(x_5)$ is negative, the odd order root would be found without too much trouble. But the even order root between x_1 and x_2 would be lost, since $f(x)$ has the same sign on both sides of the root.

The solution to the problem is to examine the derivative $f'(x)$ at each x_n to look for a change in sign. If the derivative changes sign this implies a maximum or a minimum somewhere between two points x_n and x_{n-1}, and we then have to examine the function further to look for possible roots.

If this sounds a bit haphazard, it is. That is, everything depends on the proper choice of step size h, and we are always torn between the desire to make it small for the sake of accuracy, and large for the sake of speed and low cost.

Rather than jump immediately into a brute force search program, it often is wiser to manually analyze the function and try to see, from physical reasoning, where its roots must be.

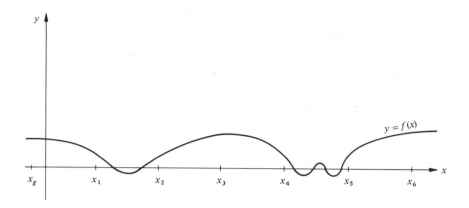

a. Closely spaced roots.

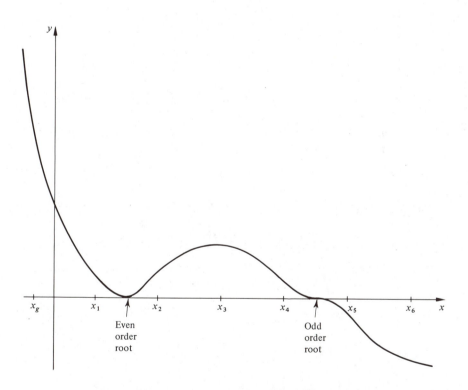

b. Even and odd order multiple roots.

Figure 3-8.   Difficult-to-find roots.

3.13 Example 3-5. Brute Force Search for a Root

As shown in Figure 3-9, the function $f(x) = x^3 - 11x^2 + 39x - 45 = 0$ has a double root at $x = 3$, and a single root at $x = 5$. Given only the information that the roots are between -10 and $+10$, however, let us write a program which will approximately find the location of the roots.

We first must decide on the step size. Let us agree on using 200 steps of size 0.1; this will take us from $x = -10$ to $x = +10$.

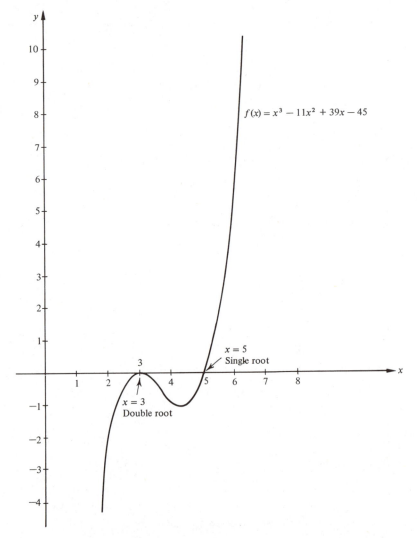

Figure 3-9.   Graph of the function $f(x) = x^3 - 11x^2 + 39x - 45$.

At each step we shall calculate $f(x)$, and also the derivative

$$f'(x) = 3x^2 - 22x + 39.$$

Our criteria for indicating a possible root shall be either: (1) $f(x)$ either becomes zero or (more likely) changes sign between two successive steps, indicating a crossover; or (2) $f'(x)$ either becomes zero or changes sign between two successive steps, *and* $f(x)$ is very small. A zero derivative merely indicates a maximum or minimum in the curve, and is not indicative of a root unless this minimum or maximum occurs on the x axis or very near it.

All the calculations in the following program are contained inside a DO loop which counts off exactly 200 repetitions. During each repetition we calculate the function $f(x_n)$, which is called FNEW in the program, and the derivative $f'(x_n)$, which is called DNEW. The previous values of $f(x_{n-1})$ and $f'(x_{n-1})$, calculated during the previous repetition of the loop, are called FOLD and DOLD.

First in the loop we calculate the product FOLD·FNEW. If the product is negative we have just passed a root and so we print "root is near"; if the product is zero then either FNEW is zero (indicating we have accidentally stumbled on a root), or FOLD is zero (which is ignored since it would have been detected in the previous repetition), or else there is an underflow and the product is zero although neither number is zero; this may indicate a close root.

If, on the other hand, the product FNEW·FOLD is positive, we then check the product DOLD·DNEW. If the product is positive then no change of derivative exists, and we go on. If on the other hand the product is negative or zero there might be a maximum or minimum in the function, and so we check whether FNEW, the value of the function $f(x_n)$ is less than an arbitrary constant Z in magnitude; that is, whether the maximum or minimum is close to the x axis. If so, then there may be a possible root.

Finally, at the end of each loop we set

FOLD = FNEW

DOLD = DNEW

to have the old values of the function and derivatives during the next repetition of the loop.

We now write the program for the function

$$f(x) = x^3 - 11x^2 + 39x - 45$$
$$f'(x) = 3x^2 - 22x + 39.$$

```
F(X) = X**3 - 11. * X**2 + 39. * X - 45.
D(X) = 3. * X  **2 - 22. * X + 39.
H = 0.1
XG = -10.
Z = 0.01
FOLD = F(XG)
DOLD = D(XG)
DO 1000 N = 1, 200
X = XG + N * H
FNEW = F(X)
DNEW = D(X)
IF (FNEW * FOLD) 10, 30, 70
10 WRITE (3, 20) X
20 FORMAT (' ROOT IS NEAR ', F15.8)
GO TO 900
30 IF (FOLD) 40, 900, 40
40 IF (FNEW) 10, 50, 10
50 WRITE (3, 60) X
60 FORMAT (' ROOT IS AT ', F15.8)
GO TO 900
70 IF (DNEW * DOLD) 80, 80, 900
80 IF (ABS(FNEW) - Z) 90, 90, 900
90 WRITE (3, 100) X, FNEW
100 FORMAT (' POSSIBLE ROOTS AT ', F15.8, ' FNEW IS ', E15.8)
900 FOLD = FNEW
1000 DOLD = DNEW
CALL EXIT
END
```

When we run the program we get the strange result

```
ROOT IS NEAR      2.99999952
ROOT IS NEAR      3.09999943
ROOT IS NEAR      5.09999943
```

Roundoff is causing strange results indeed. First of all, we note that X is calculated by the statement

$$X = XG + N * H \text{ [which equals } -10.0 + n(0.1)]$$

and should therefore go up only in steps of 0.1 whereas the x values printed out are 2.99999952 and so on; obviously the result of roundoff error.

Another strange result is the apparent fact that all three roots seem to be separate, whereas we know that there are two identical roots at $x = 3$, and there is no root $x = 3.09999943$ at all. Why?

The reason is that the double root at $x = 3$ is being picked up by the program twice, again because of roundoff error. At $x = 2.99999952$ a slight roundoff is making $f(x)$ positive, whereas it should be just slightly negative. As a result, the program decides it has just stepped over a root since $f(x)$ has changed sign, and prints a message that a "root is near 2.99999952". The next step, at $x = 3.09999943$, again yields a negative $f(x)$, and so the program decides it has again stepped over a root and prints another message. Thus the slight error in calculating $f(x)$ at $x = 2.99999952$ causes two sign changes in $f(x)$, leading the program into printing two messages.

There is also an error in the third root, which is at $x = 5.0$ instead of $x = 5.09999943$; the reason here is the large step size. The program cannot sense a root unless it either happens to hit it very closely, or else it steps over it. Because of the large steps, it reaches almost 5.1 before it can print out that a root has been found.

If we run the same program in double precision arithmetic, the roundoff errors are so much smaller that we get the correct answer

```
ROOT IS AT      3.00000000
ROOT IS AT      5.00000000
```

3.14 Roots of Polynomial Equations by Synthetic Division

The methods in this chapter will work for both polynomial equations and other types of equations; since the roots of polynomial equations crop up so often, however, it pays to look at these in further detail.

The equations we shall work with are of the type

$$f(x) = a_n x^n + a_{n-1} x^{n-1} + \cdots + a_2 x^2 + a_1 x + a_0 = 0.$$

Since we often need to evaluate the value of $f(x)$, and sometimes also of $f'(x)$ as in the Newton-Raphson method, we would like to develop a possibly more efficient way of evaluating polynomials. Since $f'(x)$ is also a polynomial of the form

$$f'(x) = n a_n x^{n-1} + (n-1) a_{n-1} x^{n-2} + \cdots + 2 a_2 x + a_1,$$

this will be useful in evaluating both polynomial functions and their derivatives.

The method to be discussed is Horner's rule, already briefly introduced in Chapter 2.

Suppose we have a polynomial

$$f(x) = 2x^3 - 5x^2 + 3x - 1,$$

which we would like to evaluate. In its present form, we would need 2 multiplications to get x^3 and x^2, and 3 multiplications to multiply each power of x by its coefficient, as well as three additions or subtractions. In general, if we have a polynomial of degree n, we need $(2n - 1)$ multiplications and n additions or subtractions.

On the other hand, we can factor this polynomial as follows:

$$\begin{aligned} f(x) &= 2x^3 - 5x^2 + 3x - 1 \\ &= [2x^2 - 5x + 3]x - 1 \\ &= [(2x - 5)x + 3]x - 1. \end{aligned}$$

As a FORTRAN statement we would write this simply as

$$F(X) = ((2. * X - 5.) * X + 3.) * X - 1.$$

which involves only 3 multiplications and 3 additions or subtractions. Hence an nth-degree polynomial now needs only n multiplications instead of $(2n - 1)$, a saving of almost 50 percent. Since each multiplication involves a roundoff error, this procedure is also likely to be more accurate, as we saw in Chapter 2.

But Horner's rule has a few other by-products which can be useful, especially in hand computation (their saving in computer time is smaller).

A technique often taught in high school, called *synthetic division*, happens to be a version of Horner's rule. As an example, suppose we wish to evaluate the function $f(x) = 2x^3 - 5x^2 + 3x - 1$ at a value $x = 2$. Synthetic division proceeds as follows:

$$
\begin{array}{r|rrrr}
2 & 2 & -5 & 3 & -1 \\
  &   & 4 & -2 & 2 \\
\hline
  & 2 & -1 & 1 & \boxed{1}
\end{array}
$$

As a review, we do synthetic division like this. Write down just the coefficients of the function of x:

$$2 \quad -5 \quad 3 \quad -1$$

To the left put the value of x for which we are evaluating the function, placing this value in a box $\rule{1em}{0.4pt}|$:

$$
\begin{array}{r|rrrr}
2 & 2 & -5 & 3 & -1
\end{array}
$$

Now draw a line under the coefficients, and bring down the first coefficient under the line:

$$
\begin{array}{r|rrrr}
2 & 2 & -5 & 3 & -1 \\
\hline
  & 2 &   &   &
\end{array}
$$

Multiply the value of x by the first coefficient, and place under the second

coefficient; then add to the second coefficient and place the sum under the
line:

$$
\begin{array}{r|rrrr}
2 & 2 & -5 & 3 & -1 \\
  &   & 4 &   &   \\
\hline
  & 2 & -1 &   &
\end{array}
$$

Now repeat the same step for the next two columns:

$$
\begin{array}{r|rrrr}
2 & 2 & -5 & 3 & -1 \\
  &   & 4 & -2 & 2 \\
\hline
  & 2 & -1 & 1 & \boxed{1}
\end{array}
$$

The last number, again boxed off, is the value of the function.

Suppose we write down each of the multiplications and additions as
follows:

$$
\begin{aligned}
2 \times 2 &= 4 \\
-5 + 4 &= -1 \\
-1 \times 2 &= -2 \\
3 - 2 &= 1 \\
1 \times 2 &= 2 \\
-1 + 2 &= 1 = \text{the value of the function.}
\end{aligned}
$$

In the same way, let us write down the multiplications and additions (or
subtractions) done by the computer in evaluating the FORTRAN statement

$$
F(X) = ((2. * X - 5.) * X + 3.) * X - 1,
$$

with $x = 2$. Since the evaluation is done left-to-right, inner parentheses first,
we get

$$
\begin{aligned}
2 \times 2 &= 4 \\
4 - 5 &= -1 \\
-1 \times 2 &= -2 \\
-2 + 3 &= 1 \\
1 \times 2 &= 2 \\
2 - 1 &= 1 = \text{the value of the function.}
\end{aligned}
$$

This sequence of steps is identical with that of synthetic division, except for the reversal of the additions and subtractions. Hence synthetic division is really Horner's rule.

To generalize somewhat, let us consider a 3rd-degree polynomial

$$f(x) = a_3 x^3 + a_2 x^2 + a_1 x + a_0,$$

which we evaluate for a particular value of x called x_n.

We start synthetic division as follows:

$$\underline{x_n|} \quad a_3 \quad a_2 \quad a_1 \quad a_0$$

Carrying through the division with symbols, we have

$$
\begin{array}{c|c|c|c|c}
\underline{x_n|} & a_3 & a_2 & a_1 & a_0 \\
 & & a_3 x_n & a_2 x_n + a_3 x_n^2 & a_1 x_n + a_2 x_n^2 + a_3 x_n^3 \\
\hline
 & a_3 & a_2 + a_3 x_n & a_1 + a_2 x_n + a_3 x_n^2 & a_0 + a_1 x_n + a_2 x_n^2 + a_3 x_n^3 \\
 & \underbrace{}_{b_3} & \underbrace{}_{b_2} & \underbrace{}_{b_1} & \underbrace{\hspace{3cm}}_{b_0}
\end{array}
$$

Thus the last term on the bottom is the function evaluated at $x = x_n$. In the same way we could generalize for any degree polynomial, and find that the last term is the value of the function at the particular value x_n. Synthetic division therefore represents a very easy way of manually evaluating a function $f(x_n)$, just as the factoring of the polynomial into a Horner's rule form represents a more efficient way of computer calculation.

Returning to the above synthetic division, let us define

$$
\begin{aligned}
b_3 &= a_3 \\
b_2 &= a_2 + a_3 x_n = a_2 + b_3 x_n \\
b_1 &= a_1 + a_2 x_n + a_3 x_n^2 = a_1 + x_n(a_2 + a_3 x_n) = a_1 + b_2 x_n \\
b_0 &= a_0 + a_1 x_n + a_2 x_n^2 + a_3 x_n^3 \\
    &= a_0 + x_n(a_1 + a_2 x_n + a_3 x_n^2) = a_0 + b_1 x_n.
\end{aligned}
$$

In the general case, suppose we are working on a polynomial of degree m, which therefore has coefficients $a_m, a_{m-1}, a_{m-2}, \ldots, a_1, a_0$; we can define a set of numbers b_i such that the highest ordered b_i is

$$b_m = a_m,$$

and each of the lower-ordered b_i can be defined in terms of the next higher ordered as

$$b_i = a_i + b_{i+1}x_n, \qquad i = (m-1), (m-2), \ldots 3, 2, 1, 0.$$

Then b_0, the last b, is the value of the function at $x = x_n$.

EXAMPLE: $f(x) = 2x^3 - 5x^2 + 3x - 1$; evaluate this function at $x_n = 2$. This is essentially the case we had before, where

$$a_3 = 2$$
$$a_2 = -5$$
$$a_1 = 3$$
$$a_0 = -1.$$

This is an mth-degree polynomial in x, so that $m = 3$, and

$$b_m = b_3 = a_3 = 2$$
$$b_2 = a_2 + b_3 x_n = -5 + (2)(2) = -1$$
$$b_1 = a_1 + b_2 x_n = 3 + (-1)(2) = 1$$
$$b_0 = a_0 + b_1 x_n = -1 + (1)(2) = 1 = f(2).$$

This is merely synthetic division in a new form, where the numbers b_i represent the last line in the synthetic division

$$
\begin{array}{r|rrrr}
2 & 2 & -5 & 3 & -1 \\
  &   & 4 & -2 & 2 \\
\hline
  & 2 & -1 & 1 & 1 \\
  & b_3 & b_2 & b_1 & b_0.
\end{array}
$$

Now, the name "synthetic division" comes from the fact that the numbers b_3, b_2, b_1 (or b_m, mb_{m-1}, $\ldots$ b_2, b_1 in the general case) represent the coefficients of another polynomial which is the quotient of $f(x)$ divided by $x - x_n$.

We can show this easily enough by actually performing the following long division:

$$
\begin{array}{r}
a_3 x^2 \\
x - x_n \overline{) a_3 x^3 \qquad\qquad + a_2 x^2 \qquad\qquad + a_1 x \qquad\qquad + a_0} \\
\underline{a_3 x^3 - a_3 x_n x^2} \\
\underline{a_3 x_n x^2 + a_2 x^2} \\
(a_2 + a_3 x_n) x^2
\end{array}
$$

At this point we stop and substitute b_3 and b_2 :

$$b_3 = a_3$$
$$b_2 = a_2 + b_3 x_n$$

and
$$b_2 x^2 = (a_2 + a_3 x_n)x^2 ;$$

$$
\begin{array}{r}
b_3 x^2 \qquad\qquad + b_2 x \qquad\qquad\qquad + b_1 \\
\hline
x - x_n \, \overline{\bigl)\, a_3 x^3 \qquad\qquad + a_2 x^2 \qquad\qquad\qquad + a_1 x \qquad\qquad\qquad + a_0} \\
a_3 x^3 - b_3 x_n x^2 \\
\hline
b_2 x^2 \\
b_2 x^2 - b_2 x_n x \\
\hline
(a_1 + b_2 x_n)x \\
\underbrace{\quad\quad} \\
b_1 \qquad x \\
b_1 \qquad x - b_1 x_n \\
\hline
a_0 + b_1 x_n \\
\underbrace{\quad\quad} \\
\text{Remainder } b_0
\end{array}
$$

We see here that b_3, b_2, and b_1 are the coefficients of the quotient. In the general case, the quotient is a function $h(x)$ equal to

$$h(x) = b_m x^{m-1} + b_{m-1} x^{m-2} + \cdots + b_2 x + b_1,$$

which is only of $(m - 1)$ degree, one less than the original function. The quantity b_0 is not part of the quotient, but actually represents the remainder.

The remainder can be made part of the quotient by dividing it by the divisor. That is, if we divide a function $f(x)$ by $(x - x_n)$ we obtain

$$\frac{f(x)}{(x - x_n)} = h(x) + \frac{b_0}{x - x_n}.$$

Returning to the above example, we have

$$h(x) = b_3 x^2 + b_2 x + b_1$$
$$= 2x^2 - x + 1,$$

and if we check, we find that

$$\frac{2x^3 - 5x^2 + 3x - 1}{x - 2} = \underbrace{2x^2 - x + 1}_{\text{quotient}} + \underbrace{\frac{1}{x - 2}}_{\text{remainder}}.$$

Multiplying both sides of the general equation by $(x - x_n)$ we have

$$f(x) = (x - x_n)h(x) + b_0.$$

Differentiating,

$$f'(x) = (x - x_n)h'(x) + h(x).$$

Now suppose we need the derivative of the function at a specific value $x = x_n$; then $(x - x_n)$ equals zero, and

$$f'(x_n) = h(x_n);$$

to find $h(x_n)$ we merely repeat the synthetic division a second time to evaluate $h(x)$ at $x = x_n$.

EXAMPLE: Suppose that we are given

$$f(x) = 2x^3 - 5x^2 + 3x - 1 = 0.$$

Suppose we wish to use the Newton-Raphson method to get a new approximation from an initial guess x_0. Then

$$x_1 = x_0 - \frac{f(x_0)}{f'(x_0)}.$$

If the initial guess is $x_0 = 2$, we wish to evaluate both $f(2)$ and $f'(2)$ to use the above formula. This is easily done by two synthetic divisions:

$$
\begin{array}{r|rrrr}
2 & 2 & -5 & 3 & -1 \\
  &   & 4 & -2 & 2 \\
\hline
  & 2 & -1 & 1 & \boxed{1} = f(2)
\end{array}
$$

$$
\begin{array}{r|rrr}
2 & 2 & -1 & 1 \\
  &   & 4 & 6 \\
\hline
  & 2 & 3 & \boxed{7} = f'(2),
\end{array}
$$

and therefore

$$x_1 = x_0 - \frac{f(x_0)}{f'(x_0)} = 2 - \frac{1}{7} = 1.857$$

The use of synthetic division is therefore quite handy in evaluating functions and their derivatives for use with manual calculations of the Newton-Raphson method. We shall see another use for it in a moment.

3.15 The Problem of Multiple Roots

Quite often it happens that an equation

$$f(x) = 0$$

has several identical roots; that is, some of the roots are of higher order. Thus, for example, the equation

$$f(x) = x^3 - 11x^2 + 39x - 45 = 0$$

has a double root at $x = 3$, and a single root at $x = 5$. We could write this function as

$$f(x) = (x - 3)(x - 3)(x - 5)$$

which, if multiplied out, gives the original function. This function is shown in Figure 3-9.

Suppose that the brute force search method illustrated in Example 3-5 has revealed to us that there may be a root near $x = 3$, and another root near $x = 5$. The root near $x = 5$ can easily be found by the Newton-Raphson method, by the method of false position, or by the bisection method, and so we need not worry longer about it.

But the double root at $x = 3$ is not so easy. Since $f(x)$ on both sides of this double root is negative, we cannot use the methods of false position or bisection, and so we are forced to use the Newton-Raphson method.

When we try to iterate

$$x_{i+1} = x_i - \frac{f(x_i)}{f'(x_i)},$$

we find that the derivative approaches zero as we get closer and closer to the correct root $\bar{x} = 3$. In fact,

$$f(x) = x^3 - 11x^2 + 39x - 45$$
$$f'(x) = 3x^2 - 22x + 39$$
$$f'(3) = 27 - 66 + 39 = 0,$$

as we can see from Figure 3-9. Unless we therefore watch the program we shall get attempted division by zero when x_i comes very close to $\bar{x}$.

To see what happens, let us evaluate the quantity $f(x)/f'(x)$ at the root $\bar{x} = 3$:

$$\frac{f(\bar{x})}{f'(\bar{x})} = \frac{\bar{x}^3 - 11\bar{x}^2 + 39\bar{x} - 45}{3\bar{x}^2 - 22\bar{x} + 39} = \frac{0}{0} = ?.$$

Hence as x approaches $\bar{x} = 3$, this term approaches 0/0, and we may have difficulty evaluating it during the Newton-Raphson iteration.

The question is: does this quantity approach any particular value as x approaches 3? We can use L'Hopital's rule:

$$\lim_{x \to 3} \frac{f(x)}{f'(x)} = \lim_{x \to 3} \frac{f'(x)}{f''(x)} = \frac{0}{f''(x)} = 0,$$

since the second derivative for a double root is nonzero. This analysis is essentially similar to our analysis of the order of convergence of the Newton-Raphson method.

Since the limit of the $f(x)/f'(x)$ term is zero, this means that the numerator must go to zero faster than the denominator. Assuming no roundoff error in the computer, the quantity $f(x)$ would become zero before $f'(x)$ does.

In using the Newton-Raphson method for multiple roots we must therefore be careful to stop the computer before the value of $f'(x)$ as calculated by the computer becomes zero; otherwise an attempted division by zero will cause problems. This requires a check with an IF statement to stop the iteration as soon as $f(x)$ is zero; since $f(x)$ becomes zero before $f'(x)$ does, this is certain to stop the iteration in time (*if* no roundoff occurs; checking $f'(x)$ for zero is thus advisable too). In any case, when $f(x)$ is zero the next x_{i+1} will be the same as the present x_i and we might as well stop.

Assuming, now, that we have done all this and found the root at $x = 3$, are we done? Not really, since this is a double root. To be more specific, in the general case any root we find *may* be a multiple root. How do we tell whether a given root is a single root or a multiple root?

Just looking at an equation is no help. If the equation is a 7th-degree polynomial, for example, it is very difficult to guess where the roots might be. When we find a multiple root and think it is only a single root, we may waste time looking for other roots when in fact there are no more to be found. Moreover, if there are two roots very close to each other, we may keep finding one and miss the other altogether.

The solution is to eliminate each root as soon as we find it. To be specific, any polynomial of degree n has n roots, and can be written as

$$f(x) = a_n x^n + a_{n-1} x^{n-1} + \cdots + a_2 x^2 + a_1 x + a_0,$$

which can be factored into the form

$$f(x) = a_n(x - x_1)(x - x_2)(x - x_3) \cdots (x - x_n),$$

where the $x_1 \ldots x_n$ are the n roots. In the case of multiple roots several of these x_i can equal each other.

To eliminate a root from the function $f(x)$, we merely divide the function by $(x - x_i)$ where x_i is that root.

EXAMPLE: To eliminate the root $x = 3$ from

$$f(x) = x^3 - 11x^2 + 39x - 45,$$

we use synthetic division:

$$
\begin{array}{r|rrrr}
3 & 1 & -11 & 39 & -45 \\
  &   & 3 & -24 & 45 \\
\hline
  & 1 & -8 & 15 & 0.
\end{array}
$$

The remainder is 0, indicating that $x - 3$ went into $f(x)$ evenly, and that $f(3) = 0$. The quotient is

$$\frac{f(x)}{x - 3} = x^2 - 8x + 15$$

$$= (x - 3)(x - 5),$$

which still has another root $x = 3$, and the root $x = 5$.

The procedure is therefore to divide out any root we find, and then look for more roots. If we find the same root again, then that root must have been a multiple-order one, and we again divide it out, and proceed.

All these extra manipulations, however, increase roundoff error, so that the last few roots we find may not be exactly correct due to accumulated roundoff. We may want to check these roots against the original equation and iterate them a few times to make sure we eliminate roundoff caused by dividing out previous roots. Another way to reduce the error is to find the smallest roots first.

The synthetic division by an already found root to eliminate that root from the polynomial equation is called *deflating* the original function, and can often be done by the computer using the following algorithm (already presented in the previous section):

Suppose an mth-degree polynomial

$$f(x) = a_m x^m + a_{m-1} x^{m-1} + \cdots + a_2 x^2 + a_1 x + a_0$$

is divided by $(x - x_n)$ to give the quotient polynomial

$$h(x) = b_m x^{m-1} + b_{m-1} x^{m-2} + \cdots + b_2 x + b_1$$

and a remainder $b_0 = f(x_n)$, we may get the coefficients b_i from the recursion formula

$$b_m = a_m$$
$$b_i = a_i + b_{i+1} x_n, \qquad i = (m - 1), (m - 2), \ldots 2, 1, 0.$$

These coefficients not only give us the deflated polynomial, but also give the value of the function. Moreover, by repeating the process on this new function, we can get $f'(x_n)$, needed for the Newton-Raphson method.

3.16 Example 3-6. Finding Multiple Order Real Roots

The fourth-degree polynomial

$$f(x) = x^4 - 10x^3 + 32x^2 - 38x + 15 = 0$$

has a double root at $x = 1$, and single roots at $x = 3$ and $x = 5$; it can be expressed as

$$f(x) = (x - 1)(x - 1)(x - 3)(x - 5) = 0.$$

We wish to find the roots by the Newton-Raphson method, starting with the lowest first. We will do this by assuming $x_0 = 0$ as an initial guess, iterating until we find a root (hopefully the lowest first), and then deflating the polynomial by dividing by that root. Then we begin again with a guess $x_0 = 0$ and repeat.

First we shall need a synthetic division procedure, which we write as a separate FORTRAN subroutine to be called later in the main program:

```
SUBROUTINE SYDIV (A, B, N, X)
DIMENSION A(5), B(5)
B(1) = A(1)
DO 10 I = 1, N
10 B(I+1) = A(I+1) + B(I) * X
RETURN
END
```

This synthetic division subroutine takes an array A which can hold up to 5 coefficients (such as a_4, a_3, a_2, a_1, and a_0) in *reverse order*—the highest coefficient is in A(1) and so on—and converts it into an array B of up to 5 new coefficients (such as b_4 through b_0) from the recursion relation

$$b_m = a_m$$
$$b_i = a_i + b_{i+1}x_n, \qquad i = (m-1), (m-2), \ldots 2, 1, 0.$$

In the subroutine, the coefficients are stored backward, so that we actually start by making B(1) equal to A(1), and then use the recursion relation to find the others. The inputs to the subroutine are the arrays A and B, the number N giving the degree of the polynomial (since the deflated polynomials are of lower degree), and the value X giving the x_n to be used.

The program below is annotated and shows the procedure followed: we use the Newton-Raphson method to find a root, deflate the polynomial,

```
      DIMENSION COEFF (5), A(5), B(5)
C     INITIALIZE
      COEFF (1) = 1.
      COEFF (2) = -10.
      COEFF (3) = 32.
      COEFF (4) = -38.
      COEFF (5) = 15.
      N = 4
      NMBR = N - 1
C     FIND FIRST N-1 ROOTS
      DO 1000 J = 1, NMBR
      N1 = N - 1
      X = 0.
C     ITERATE NEWTON-RAPHSON METHOD FOR THE J-TH ROOT
      DO 100 K = 1, 1000
C     FIND F(X) AT X
      CALL SYDIV (COEFF, A, N,  X)
      FX = A(N  + 1)
C     FIND DERIVATIVE AT X
      CALL SYDIV (A, B, N1, X)
      DFX = B (N1 + 1)
C     CHECK IF SOLUTION IS FOUND
      IF (FX) 10, 300, 10
   10 IF (DFX) 20, 200, 20
   20 XNEW = X - FX / DFX
      IF (XNEW - X) 100, 300, 100
  100 X = XNEW
      WRITE (3, 150)
  150 FORMAT (' NO ROOT FOUND AFTER 1000 ITERATIONS - DIVERGES')
      CALL EXIT
  200 WRITE (3, 250)
  250 FORMAT (' DERIVATIVE IS 0 - REPEAT WITH NEW STARTING VALUE')
      CALL EXIT
  300 WRITE (3, 350) J, X
  350 FORMAT (' ROOT NO.', I5, ' IS', F15.8)
C     DEFLATE EQUATION
      CALL SYDIV (COEFF, COEFF, N,  X)
C     REDUCE DEGREE OF POLYNOMIAL BY 1
      N = N - 1
 1000 CONTINUE
C     AFTER N-1 ROOTS HAVE BEEN FOUND, THE REDUCED POLYNOMIAL  IS OF THE
C     FORM F(X) = A * X + B = 0, AND X IS THEN EQUAL TO -A/B
      X = -COEFF(2) / COEFF (1)
      WRITE (3, 350) J, X
      CALL EXIT
      END
```

and repeat. We only find the first 3 roots in this way, since after the third root the deflated polynomial is of the form $ax + b = 0$, $x = -b/a$.

This program uses the synthetic division subroutine SYDIV three times: once to evaluate $f(x)$ and once to evaluate $f'(x)$ during each repetition of the Newton-Raphson iteration which ends on statement number 100, and once to deflate the original polynomial after each of the first three roots is found.

The first part of the program initializes the constants needed: the five coefficients of the polynomial, and N, which gives the degree of the poly- nomial (and the number of roots to be found).

We obtain the following results from this program for the polynomial

$$f(x) = x^4 - 10x^3 + 32x^2 - 38x + 15 = 0.$$

```
ROOT NO.    1 IS      0.99999721
ROOT NO.    2 IS      1.00000279
ROOT NO.    3 IS      2.99999998
ROOT NO.    4 IS      5.00000002
```

Note that the results are not spectacularly accurate, mainly due to the round- off errors in the synthetic division subroutine. Except for this roundoff error, however, we have managed to find all four roots of the polynomial, and could use straightforward double-precision Newton-Raphson iteration to improve the accuracy; the single-order roots (or odd-order multiple roots) could also be obtained from the method of false position or the bisection method. For best accuracy we would work on the original polynomial with- out deflating it once we have found the approximate roots.

3.17 Complex Roots of Equations by the Newton- Raphson Method

So far in this chapter we have been discussing equations with real co- efficients and real roots. Quite often, however, we encounter a polynomial equation with real coefficients, but whose roots are complex.

Generally such an equation can be solved by the Newton-Raphson method merely by assuming an initial guess x_0 as complex.

Suppose we have

$$f(x) = a_n x^n + a_{n-1} x^{n-1} + \cdots + a_2 x^2 + a_1 x + a_0,$$

where the coefficients a_i are all real. Hence $f'(x)$ is a polynomial of degree $(n - 1)$, and its coefficients are also real:

$$f'(x) = na_n x^{n-1} + (n - 1)a_{n-1} x^{n-2} + \cdots + 2a_2 x + a_1.$$

If the initial guess x_0 is real, then

$$x_1 = x_0 - \frac{f(x_0)}{f'(x_0)}$$

will also be real, and in fact all the x_i will be real; we can therefore never find a complex root if we start with a real x_0.

If, on the other hand, the initial guess x_0 is complex, then x_1 will also be complex, and so on. We may therefore find the actual complex root $\bar{x}$, *assuming the process converges.* Just as some real roots cannot be found by this method because of nonconvergence, so we sometimes cannot find complex roots and must resort to other methods, and so there is no guarantee that this will work. But the Newton-Raphson method works so fast (when it works at all) that it is well worth a try.

Several methods can be used in addition to simple Newton-Raphson iteration. In general, a polynomial with real coefficients will always have an *even* number of complex roots, such that for every root there is also its complex conjugate; this is required to make the coefficients real. And so as soon as we find a root $x_1 = \alpha + j\beta$ (where $j = \sqrt{-1}$), we also know that another root is $x_2 = \alpha - j\beta$.

Since the terms $(x - x_1)$ and $(x - x_2)$ are both factors of the original polynomial, so is their product. When we multiply $(x - x_1) = x - \alpha - j\beta$ by the factor $(x - x_2) = x - \alpha + j\beta$, we get $x^2 - 2\alpha x + (\alpha^2 + \beta^2)$. Dividing this factor out of the original polynomial we deflate its degree by 2, and then start again.

Further, the original polynomial may have some real roots which can be found by the methods described earlier in this chapter; we can then deflate the polynomial by removing these roots, and work on the simplified polynomial.

3.18 Example 3-7. Finding a Complex Root by the Newton-Raphson Method

As a simple example, let us find the complex roots of

$$f(x) = x^2 + x + 1 = 0.$$

Since the coefficients are real and the polynomial is of the second degree, there are two roots and they must be complex conjugates of each other. We therefore only need to find one root.

From the quadratic equation, we know that the roots are

$$x_{1,2} = \frac{-b \pm \sqrt{b^2 - 4ac}}{2a} = \frac{-1 \pm \sqrt{1 - 4}}{2} = -\frac{1}{2} \pm j\frac{\sqrt{3}}{2}$$

$$= -0.5 \pm j0.866,$$

where $j = \sqrt{-1}$.

From the Newton-Raphson method,

$$x_{i+1} = x_i - \frac{f(x_i)}{f'(x_i)},$$

where

$$f(x) = x^2 + x + 1$$
$$f'(x) = 2x + 1,$$

and

$$x_{i+1} = x_i - \frac{x_i^2 + x_i + 1}{2x_i + 1}.$$

However all of these calculations must be done with complex arithmetic, and only a few FORTRAN systems (those employing the full FORTRAN IV on a larger computer) are equipped for complex arithmetic. In this example we will show how to get around this by storing the real (RE) and imaginary (IM) parts of each number separately, and calculating the real and imaginary parts in separate steps.

To perform the Newton-Raphson iteration, we must start with a complex guess x_0. Suppose we assume that $x_0 = 1 + j$. In FORTRAN, we write

```
XRE = 1.
XIM = 1.
```

We must now calculate the function $f(x_0) = x_0^2 + x_0 + 1$, and so we first calculate x_0^2:

$$(1 + j)(1 + j) = (1 - 1) + (j + j) = 2j.$$

In FORTRAN, we write

```
X2RE = XRE * XRE - XIM * XIM
X2IM = 2. * XRE * XIM
```

Now, $f(x_0) = x_0^2 + x_0 + 1 = 2j + 1 + j + 1 = 2 + 3j$. In FORTRAN we add the real and imaginary parts separately:

    FXRE = X2RE + XRE + 1.
    FXIM = X2IM + XIM

Further, we need

$$f'(x_0) = 2x_0 + 1 = 2(1 + j) + 1$$
$$= 2 + 2j + 1 = 3 + 2j.$$

In FORTRAN:

    DFXRE = 2. * XRE + 1.
    DFXIM = 2. * XIM

Now we solve for x_1:

$$x_1 = x_0 - \frac{f(x_0)}{f'(x_0)} = (1 + j) - \frac{2 + 3j}{3 + 2j}.$$

To evaluate the fraction we may multiply the top and bottom by $(3 - 2j)$, the complex conjugate of the denominator; this will make the denominator real.

$$\frac{2 + 3j}{3 + 2j} \cdot \frac{3 - 2j}{3 - 2j}.$$

Multiplying out the numerator we have

$$(2 + 3j)(3 - 2j) = (6 + 6) + j(9 - 4) = 12 + 5j.$$

In FORTRAN the same calculation becomes

    TOPRE = FXRE * DFXRE + FXIM * DFXIM
    TOPIM = FXIM * DFXRE - FXRE * DFXIM

Similarly, multiplying out the denominator gives us

$$(3 + 2j)(3 - 2j) = 9 + 6j - 6j + 4 = 9 + 4 = 13,$$

which is entirely real. In FORTRAN, we write this as

    BOT = DFXRE * DFXRE + DFXIM * DFXIM

Now we have

$$x_1 = (1 + j) - \frac{12 + 5j}{13}$$

$$= 1 + j - \tfrac{12}{13} - \tfrac{5}{13}j = \tfrac{1}{13} + \tfrac{8}{13}j = 0.77 + 0.62j.$$

In FORTRAN the old real part of x is replaced by the newly calculated real part

$$XRE = XRE - TOPRE / BOT$$

and the imaginary part is also replaced by the newly calculated

$$XIM = XIM - TOPIM / BOT$$

This essentially completes the complex arithmetic calculations, except for the addition of output statements, and gives us the following FORTRAN program:

```
        XRE = 1.
        XIM = 1.
        WRITE (3, 10) XRE, XIM
     10 FORMAT (4X, 2F15.8, '*J')
        DO 1000 I = 1, 10
        X2RE = XRE * XRE - XIM * XIM
        X2IM = 2. * XRE * XIM
        FXRE = X2RE + XRE + 1.
        FXIM = X2IM + XIM
        DFXRE = 2. * XRE + 1.
        DFXIM = 2. * XIM
        TOPRE = FXRE * DFXRE + FXIM * DFXIM
        TOPIM = FXIM * DFXRE - FXRE * DFXIM
        BOT = DFXRE * DFXRE + DFXIM * DFXIM
        XRE = XRE - TOPRE / BOT
        XIM = XIM - TOPIM / BOT
        WRITE (3, 100) I, XRE, XIM
    100 FORMAT (' X',I2, 2F15.8, '*J')
   1000 CONTINUE
        CALL EXIT
        END
```

Starting with the initial value of $1 + j$, the following printout shows us that the program provides the correct answer after only 7 iterations:

```
            1.00000000      1.00000000*J
     X 1    0.07692307      0.61538461*J
     X 2   -0.51559251      0.63201663*J
     X 3   -0.49316686      0.90898620*J
     X 4   -0.49968450      0.86701730*J
     X 5   -0.49999963      0.86602591*J
     X 6   -0.49999999      0.86602540*J
     X 7   -0.50000000      0.86602540*J
     X 8   -0.50000000      0.86602540*J
     X 9   -0.50000000      0.86602540*J
     X10   -0.50000000      0.86602540*J
```

We shall see more of complex roots, as well as solutions of simultaneous equations, in the next chapter.

3.19 Problems

1. For each of the following equations $f(x) = 0$ rearrange the expression into the form $x = g(x)$ in such a way that the iteration converges, and write the program to find a root.

  a. $x^2 - 5x + 6 = 0$,
  b. $x^2 - 6x + 9 = 0$,
  c. $\ln x - x + 2 = 0$,
  d. $x^2 - 4 = 0$.

2. The function $x^2 - x = 0$ has roots at $x = 0$ and $x = 1$. Which root can we be sure of finding with the iteration $x_{i+1} = (x_i)^2$ and a starting value $x_0 = \frac{1}{2}$? Why?

3. Another way of tackling the function $x^2 - x = 0$ is to add $10x$ to each side and divide by 10, to get

$$x = g(x) = \frac{x^2 + 9x}{10}.$$

What range of starting values x_0 will lead to a guaranteed solution? Write a program in FORTRAN to find a root.

4. For the iteration of problem 3, graph $g'(x)$ for x between -2 and $+2$. Be sure to plot several points between 0 and 1.

5. The Newton-Raphson iteration for $x^2 - x = 0$ is

$$x_{i+1} = x_i - \frac{x_i^2 - x_i}{2x_i - 1} = \frac{x_i^2}{2x_i - 1} = g(x_i).$$

Write a FORTRAN program to find the roots using starting values of -10, $+5$, and $+\frac{1}{2}$, in that order. Explain your results.

6. For the iteration of problem 5, plot the functions $g(x)$ and $g'(x)$ for $0 \leqslant x \leqslant 1$. Examine both especially for $x = 0$, $x = \frac{1}{2}$, and $x = 1$. Is the process second-order? Will it converge for a starting value of $x_0 = \frac{1}{2}$?

7. Finding the square root of a number a is the same as finding the root of $f(x) = x^2 - a = 0$. Examine the convergence of the iteration $x_{i+1} = a/x_i$, for any positive a.

8. Use the Newton-Raphson method to solve $x^2 - 2 = 0$. Compare this method with Example 1-3.

9. Develop the Newton-Raphson iterations for finding the cube root and the fourth root of a number.

10. The program of Example 3-3 is somewhat inefficient because for each iteration it calculates $f(x_L)$ and $f(x_R)$, although these have already been found in the previous iteration. Modify the program so that these values are saved from one iteration to the next. Repeat for the program of Example 3-4. Execute both programs if a computer is available.

11. After modifying the program of Example 3-3 as suggested in problem 10, make the following further modification and repeat; compare the number of iterations needed for the original and the modified programs. The modification is this: When values of $f(x_L)$ and $f(x_R)$ are used from the previous iteration, one of them comes from the previous $f(x_M)$, while the other is the previous $f(x_L)$ or $f(x_R)$ without change; change the program so that the $f(x_L)$ or $f(x_R)$ which was used from the previous iteration is divided by 2 before being used again.

12. Try the following problems by using the Newton-Raphson method, false position, and bisection; use a brute-force program to find approximate roots for the latter two methods.

a. $\ln x - x + 2 = 0$,
b. $\sin x - x^2 = 0$ (Do not find the root at $x = 0$),
c. $x - 2 \cos x = 0$,
d. $x - 2 \sin x = 0$ (Do not find the root at $x = 0$),
e. $xe^x - 2 = 0$,
f. $x \log_{10} x - 10 = 0$,
g. $\sin x - \csc x + 1 = 0$.

13. Try the Newton-Raphson method for

$$x^2 - 4x + 4.1 = 0.$$

Make sure to include a counter to stop the program after 100 iterations. Why does the method not work? Try again using complex arithmetic and a complex value for x_0.

14. Another method for finding the roots of $f(x) = 0$, called the secant method, is a cross between the method of false position and the Newton-Raphson method. It often converges almost as fast as the Newton-Raphson method, but avoids the need for calculating the derivative $f'(x)$. Instead of using a tangent line as in Figure 3-5, we use a secant line as in Figure 3-10. By analogy with Eq. (3-11), develop the iteration

equation for this method, compare with Eq. (3-14), and apply it to the
equation solved in Example 3-2.

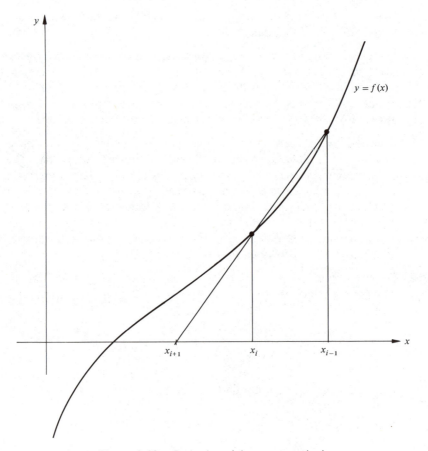

Figure 3-10.   Derivation of the secant method.

15. To solve for x in $f(x) = 0$, let

$$g(x) = x + a_0 + a_1 y + \cdots + a_n y^n,$$

where $y = f(x)$. Choose $a_0, a_1, \ldots, a_n$ in terms of $f(\bar{x})$, $f'(\bar{x})$, etc.,
so that $g(\bar{x}) = \bar{x}$, and $g'(\bar{x}) = g''(\bar{x}) = \cdots = g^{(n)}(\bar{x}) = 0$. For $n = 1$,
compare with the Newton-Raphson method. For $n = 2$, write out the
iteration formula suggested by this method (it is called the Richmond
method).

Nonlinear Simultaneous Equations

In Chapter 3 we saw how to find the roots of an algebraic equation of the form

$$f(x) = 0$$

In such an equation we had the one variable x, and had to find those values of x which satisfied the equation.

In the general case we may have several equations and several unknowns, and must find those values of the unknowns which satisfy *all* the equations at the same time. There are often several such values which work.

Because there are different ways of solving them, we distinguish between *linear* simultaneous equations, which we shall solve with matrix methods in Chapter 5, and *nonlinear* simultaneous equations which we shall study in this chapter. The methods of this chapter will work for the linear case also, but are often not as convenient as the matrix methods to come later.

4.1 Introduction to Simultaneous Equations

As a simple case of simultaneous equations consider the following:

$$3x + 2y - 5 = 0$$
$$2x + 3y - 5 = 0.$$

These are two equations in the two unknowns x and y, and we can quite simply see that $x = y = 1$ is the solution which satisfies both equations. We can write these two equations, using similar notation to that of Chapter 3, as

$$f_1(x, y) = 0$$
$$f_2(x, y) = 0,$$

and now the object is to find values of x and y which satisfy both equations.

If we examine each equation separately, we see that each equation considered by itself has an infinite number of possible solutions. Thus the first equation

$$3x + 2y - 5 = 0$$

can be satisfied if $x = 0$ and $y = 2\frac{1}{2}$, if $x = y = 1$, or if $x = 2$ and $y = -\frac{1}{2}$, or by any number of other combinations. Those combinations which satisfy this first equation can be plotted as the infinitely long line AA' in Figure 4-1.

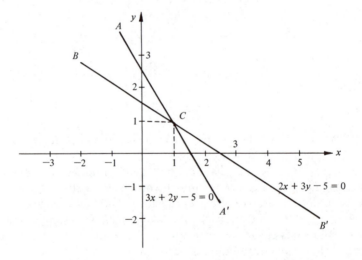

Figure 4-1. The solution of two simultaneous equations is at the intersection of the separate solutions.

In the same way any of the combinations of x and y on line BB' in Figure 4-1 will satisfy the second equation. However, only the one point C at $x = y = 1$, which lies at the intersection of the two lines, satisfies both equations simultaneously. In this case we have two equations in two unknowns and can solve them to obtain a unique answer, although this is not always the case. For example, the equations

$$3x + 2y - 5 = 0$$
$$6x + 4y - 10 = 0$$

both describe the line AA' in Figure 4-1, and any value of x and y which satisfies one equation satisfies the other; in this case there is no unique solution. We will discuss this case further in Chapter 5.

The two simultaneous equations plotted in Figure 4-1 are both called *linear* because (1.) the unknowns x and y appear in no higher powers than x^1 and y^1, and are never multiplied by each other, and (2.) as a result the graph of each equation is a straight line. Linear equations, unless there are very many of them with many unknowns, are fairly simple to solve (at least in theory).

Often, however, we have several simultaneous equations to be solved which are not linear. Figure 4-2 shows the curves defined by two *nonlinear* simultaneous equations in two unknowns x and y. The combined solution to the two equations consists of the values of x and y at the four intersections of the two curves. As shown in the figure, the solution to two nonlinear simultaneous equations can consist of many points, and sometimes even an

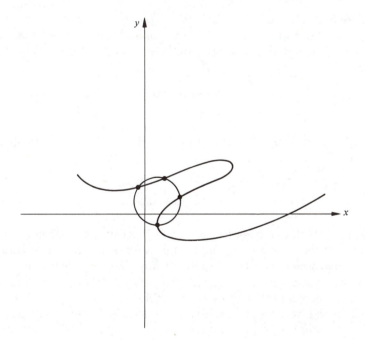

Figure 4-2.   Nonlinear simultaneous equations may have many intersections and hence many solutions.

infinite number of points. The solution of even just two nonlinear equations in two unknowns can be quite difficult.

Nonlinear equations, in addition to the fact that their individual solutions are not straight lines in two dimensions (or planes in three dimensions, and so on), are further characterized by the fact that powers of the variables appear, and that often different variables may multiply each other. For example, the following is a set of two nonlinear equations in the two unknowns x and y:

$$xy - x - 1 = 0$$
$$y - x^2 = 0.$$

4.2 Iterative Solution of Two Simultaneous Nonlinear Equations

As a general case let us consider two nonlinear equations in two unknowns, written as

$$f_1(x, y) = 0$$
$$f_2(x, y) = 0 \tag{4-1}$$

and try to find values of x and y which satisfy both equations.

Following the lead of Chapter 3, we try an iterative solution by taking an x out of the first equation, taking a y out of the second equation, and writing both as

$$x = g_1(x, y)$$
$$y = g_2(x, y).$$

As before, we shall try to obtain the $(i + 1)$th guess from the ith guess by iterating

$$x_{i+1} = g_1(x_i, y_i)$$
$$y_{i+1} = g_2(x_i, y_i). \tag{4-2}$$

We then start with initial guesses x_0 and y_0, calculate new values x_1 and y_1, and continue the same process, hoping that after each iteration the values of x_i and y_i approach the actual solutions $\bar{x}$ and $\bar{y}$. The question now is whether this process converges to the correct roots. For a proof we shall need the mean value theorem for two or more variables.

In Chapter 3 we gave the mean value theorem for a function $g(x)$ in the one variable x as

$$g(b) - g(a) = g'(\xi)(b - a)$$

for some value of ξ taken between a and b. In slightly different notation, somewhat more appropriate to this chapter, we might rewrite this same expression as

$$g(b) - g(a) = \frac{dg}{dx}_{\text{at } x=\xi} (b-a),$$

where the notation $dg/dx_{\text{at } x=\xi}$ means the derivative of the function $g(x)$ with respect to x, evaluated at the specific value $x = \xi$.

In words, we might express this theorem as follows: The change in the function $g(x)$ as x moves from a to b can be expressed as the derivative dg/dx evaluated at some ξ between a and b (which can be interpreted as "the *average* rate of change of the function $g(x)$ with a change of x"), times the actual change in x (equal to $b - a$) which caused it.

Now suppose that we have a function $g(x, y)$ having the two variables x and y, so that the value of the function g depends on both x and y.

Further assume that we start at the point $x = a$ and $y = c$, and that x and y both change by a small amount so that we arrive at the point $x = b$ and $y = d$. In going from the point (a, c) to the point (b, d), $g(x, y)$ changes by some amount

$$g(b, d) - g(a, c).$$

This change in g can be interpreted as being partially due to the fact that we changed x, and partially due to the fact we changed y; in fact, the total change is the sum of the change due to moving x plus the change due to moving y. Hence, just as in the single-variable case, we can now write a mean value theorem for functions of two variables:

Mean Value Theorem for Two Variables   *If a function $g(x, y)$ is continuous, and has the partial derivatives $\partial g/\partial x$ and $\partial g/\partial y$ which are both continuous, we can write*

$$g(b, d) - g(a, c) = \left(\frac{\partial g}{\partial x}\right)_{\substack{\text{at } x=\xi \\ y=\eta}} (b - a) + \left(\frac{\partial g}{\partial y}\right)_{\substack{\text{at } x=\lambda \\ y=\mu}} (d - c),$$

where ξ and λ are taken between a and b, while η and μ are taken between c and d.

In this theorem $\partial g/\partial x$ is the *partial* derivative of the function $g(x, y)$ with respect to x, assuming that y is constant, and evaluated at $x = \xi$ and $y = \eta$. Similarly, $\partial g/\partial y$ is the partial derivative with respect to y, assuming that x is constant, and evaluated at $x = \lambda$ and $y = \mu$. We see that the right side of the equation is just the double application of the single-variable theorem we had previously, once to get the change in $g(x, y)$ caused by the change of x, and once to get the change in $g(x, y)$ caused by the change of y.

Following the lead of Chapter 3, we shall now apply the theorem to an iteration of two simultaneous equations. If we start with initial values of x_0 and y_0 we may write

$$\begin{cases} x_1 = g_1(x_0, y_0) \\ y_1 = g_2(x_0, y_0) \end{cases}$$

$$\begin{cases} x_2 = g_1(x_1, y_1) \\ y_2 = g_2(x_1, y_1) \end{cases}$$

$$\vdots$$

$$\begin{cases} x_i = g_1(x_{i-1}, y_{i-1}) \\ y_i = g_2(x_{i-1}, y_{i-1}) \end{cases}$$

$$\begin{cases} x_{i+1} = g_1(x_i, y_i) \\ y_{i+1} = g_2(x_i, y_i) \end{cases}.$$

We now substitute into the mean value theorem, letting

$$a = x_{i-1}$$
$$b = x_i$$
$$c = y_{i-1}$$
$$d = y_i.$$

We first apply the mean value theorem to the function $g_1(x, y)$:

$$g_1(x_i, y_i) - g_1(x_{i-1}, y_{i-1})$$

$$= \left(\frac{\partial g_1}{\partial x}\right)_{\substack{\text{at } x = \xi_i^{(1)} \\ y = \eta_i^{(1)}}} (x_i - x_{i-1}) + \left(\frac{\partial g_1}{\partial y}\right)_{\substack{\text{at } x = \lambda_i^{(1)} \\ y = \mu_i^{(1)}}} (y_i - y_{i-1}), \quad (4\text{-}3)$$

where $\xi_i^{(1)}$ and $\lambda_i^{(1)}$ are between x_i and x_{i-1}, and $\eta_i^{(1)}$ and $\mu_i^{(1)}$ are between y_i and y_{i-1}; the superscript [1] indicates that these apply to g_1 and not to g_2.

We now repeat the same steps for the function $g_2(x, y)$:

$$g_2(x_i, y_i) - g_2(x_{i-1}, y_{i-1})$$

$$= \left(\frac{\partial g_2}{\partial x}\right)_{\substack{\text{at } x = \xi_i^{(2)} \\ y = \eta_i^{(2)}}} (x_i - x_{i-1}) + \left(\frac{\partial g_2}{\partial y}\right)_{\substack{\text{at } x = \lambda_i^{(2)} \\ y = \mu_i^{(2)}}} (y_i - y_{i-1}), \quad (4\text{-}4)$$

where $\xi_i^{(2)}$ and $\lambda_i^{(2)}$ are between x_i and x_{i-1}, while $\eta_i^{(2)}$ and $\mu_i^{(2)}$ are between y_i and y_{i-1}; the superscript [2] indicates that these apply to g_2 and not necessarily to g_1.

Looking at the left-hand sides of these two equations, we note that

$$x_{i+1} = g_1(x_i, y_i)$$
$$x_i = g_1(x_{i-1}, y_{i-1})$$
$$y_{i+1} = g_2(x_i, y_i)$$
$$y_i = g_2(x_{i-1}, y_{i-1}).$$

Hence Eqs. (4-3) and (4-4) can be simplified to

$$\left| x_{i+1} - x_i \right| \leqslant \left| \frac{\partial g_1}{\partial x} \right|_{\substack{\text{at } x = \xi_i^{(1)} \\ y = \eta_i^{(1)}}} \left| (x_i - x_{i-1}) \right| + \left| \frac{\partial g_1}{\partial y} \right|_{\substack{\text{at } x = \lambda_i^{(1)} \\ y = \mu_i^{(1)}}} \left| (y_i - y_{i-1}) \right|$$

$$\left| y_{i+1} - y_i \right| \leqslant \left| \frac{\partial g_2}{\partial x} \right|_{\substack{\text{at } x = \xi_i^{(2)} \\ y = \eta_i^{(2)}}} \left| (x_i - x_{i-1}) \right| + \left| \frac{\partial g_2}{\partial y} \right|_{\substack{\text{at } x = \lambda_i^{(2)} \\ y = \mu_i^{(2)}}} \left| (y_i - y_{i-1}) \right|.$$

To get these into the required form, we take absolute values of both sides, and then take absolute values of each term on the right. This introduces a ⩽ sign since the sum of two absolute values is greater than or equal to the absolute value of the sum, by the triangle inequality.

We now add the two inequalities together as follows:

$$\left| x_{i+1} - x_i \right| + \left| y_{i+1} - y_i \right| \leqslant \left[\left| \frac{\partial g_1}{\partial x} \right|_{\substack{\text{at } x = \xi_i^{(1)} \\ y = \eta_i^{(1)}}} + \left| \frac{\partial g_2}{\partial x} \right|_{\substack{\text{at } x = \xi_i^{(2)} \\ y = \eta_i^{(2)}}} \right] \left| x_i - x_{i-1} \right|$$

$$+ \left[\left| \frac{\partial g_1}{\partial y} \right|_{\substack{\text{at } x = \lambda_i^{(1)} \\ y = \mu_i^{(1)}}} + \left| \frac{\partial g_2}{\partial y} \right|_{\substack{\text{at } x = \lambda_i^{(2)} \\ y = \mu_i^{(2)}}} \right] \left| y_i - y_{i-1} \right|.$$

In each of these expressions the ξ, η, λ, and μ are different, and depend on where in the iteration we are. But suppose that in the region of interest, including all of the x_i and y_i as well as the actual roots $\bar{x}$ and $\bar{y}$, the partial derivatives are bounded. We may then write

$$\left| \frac{\partial g_1}{\partial x} \right|_{\substack{\text{at } x = \xi_i^{(1)} \\ y = \eta_i^{(1)}}} + \left| \frac{\partial g_2}{\partial x} \right|_{\substack{\text{at } x = \xi_i^{(2)} \\ y = \eta_i^{(2)}}} \leqslant M$$

$$\left| \frac{\partial g_1}{\partial y} \right|_{\substack{\text{at } x = \lambda_i^{(1)} \\ y = \mu_i^{(1)}}} + \left| \frac{\partial g_2}{\partial y} \right|_{\substack{\text{at } x = \lambda_i^{(2)} \\ y = \mu_i^{(2)}}} \leqslant M$$

for some upper bound M and for any i. If this is so, then

$$\left| x_{i+1} - x_i \right| + \left| y_{i+1} - y_i \right| \leqslant M \left| x_i - x_{i-1} \right| + M \left| y_i - y_{i-1} \right|$$
$$\leqslant M \{ \left| x_i - x_{i-1} \right| + \left| y_i - y_{i-1} \right| \}.$$

To be specific,

$$|x_2 - x_1| + |y_2 - y_1| \leqslant M\{|x_1 - x_0| + |y_1 - y_0|\}$$
$$|x_3 - x_2| + |y_3 - y_2| \leqslant M\{|x_2 - x_1| + |y_2 - y_1|\}.$$
$$\vdots$$

We now substitute the first inequality in the second, the second in the third, and so on, to obtain

$$|x_{i+1} - x_i| + |y_{i+1} - y_i| \leqslant M^i\{|x_1 - x_0| + |y_1 - y_0|\}.$$

If the iteration is to converge, the left side must approach zero for large i; a sufficient, though not necessary, condition to ensure this is that $M < 1$, so that M^i approaches zero for large i.

Hence it is sufficient to show that

$$\left|\frac{\partial g_1}{\partial x}\right| + \left|\frac{\partial g_2}{\partial x}\right| \leqslant M < 1$$

$$\left|\frac{\partial g_1}{\partial y}\right| + \left|\frac{\partial g_2}{\partial y}\right| \leqslant M < 1$$

for all x and y in the region containing all the values x_i and y_i and the correct values $\bar{x}$ and $\bar{y}$. Moreover, if M is very small in the entire region the iteration converges relatively quickly; if M is near 1 in magnitude, then the iteration may converge very slowly. This behavior is similar to the single variable case we discussed in Chapter 3.

The extension of this discussion to simultaneous equations of three and more variables is quite easy, and follows the same reasoning; it is left to the student.

4.3 The Newton-Raphson Method

In Chapter 3 we showed how the Newton-Raphson method can be derived from the Taylor series approach. We now extend this procedure to two variables; the extension to three or more variables follows the same lines.

Suppose we are solving the two simultaneous equations

$$f_1(x, y) = 0$$
$$f_2(x, y) = 0,$$

where both functions are continuous and differentiable, so that they may be expanded in a Taylor series. We use a two-dimensional Taylor series, which treats the x and y contributions separately.

We may then express $f_1(x_{i+1}, y_{i+1})$ in terms of the function and its derivatives at (x_i, y_i) as follows:

$$f_1(x_{i+1}, y_{i+1}) = f_1(x_i, y_i) + \underbrace{\frac{\partial f_1}{\partial x}(x_{i+1} - x_i)}_{\substack{\text{contribution} \\ \text{due to } x}} + \underbrace{\frac{\partial f_1}{\partial y}(y_{i+1} - y_i)}_{\substack{\text{contribution} \\ \text{due to } y}} + \cdots, \tag{4-5}$$

where the partial derivatives $\partial f_1/\partial x$ and $\partial f_1/\partial y$ are to be evaluated at $x = x_i$ and $y = y_i$.

In the same way we can expand the function f_2 as follows:

$$f_2(x_{i+1}, y_{i+1}) = f_2(x_i, y_i) + \frac{\partial f_2}{\partial x}(x_{i+1} - x_i) + \frac{\partial f_2}{\partial y}(y_{i+1} - y_i) + \cdots. \tag{4-6}$$

Now suppose that x_{i+1} and y_{i+1} are both very close to the actual root $(\bar{x}, \bar{y})$ so that the left side of the above equations is almost equal to zero; further, assume that x_i is very close to x_{i+1}, and that y_i is very close to y_{i+1}, so that the remaining terms of the Taylor series may be omitted. Then Eqs. (4-5) and (4-6) simplify to

$$0 = f_1(x_i, y_i) + \frac{\partial f_1}{\partial x}(x_{i+1} - x_i) + \frac{\partial f_1}{\partial y}(y_{i+1} - y_i)$$

$$0 = f_2(x_i, y_i) + \frac{\partial f_2}{\partial x}(x_{i+1} - x_i) + \frac{\partial f_2}{\partial y}(y_{i+1} - y_i). \tag{4-7}$$

For simplicity, let

$$x_{i+1} - x_i = h$$
$$y_{i+1} - y_i = k$$

so that we can get the $(i + 1)$th approximation from the ith by

$$x_{i+1} = x_i + h$$
$$y_{i+1} = y_i + k.$$

Our problem is now to find the values of h and k for each iteration. To do this, we substitute h and k into Eq. (4-7):

$$0 = f_1(x_i, y_i) + \frac{\partial f_1}{\partial x}h + \frac{\partial f_1}{\partial y}k$$

$$0 = f_2(x_i, y_i) + \frac{\partial f_2}{\partial x}h + \frac{\partial f_2}{\partial y}k.$$

Rearranging, we get the simultaneous equations

$$\frac{\partial f_1}{\partial x}h + \frac{\partial f_1}{\partial y}k = -f_1(x_i, y_i)$$

$$\frac{\partial f_2}{\partial x}h + \frac{\partial f_2}{\partial y}k = -f_2(x_i, y_i),$$

where the partial derivatives are evaluated at x_i and y_i.

These resulting simultaneous equations can easily be solved for h and k by determinants (see Chapter 5 for an introduction to determinants), *provided* that the determinant of the coefficients does not vanish; that is,

$$J = \begin{vmatrix} \dfrac{\partial f_1}{\partial x} & \dfrac{\partial f_1}{\partial y} \\[2mm] \dfrac{\partial f_2}{\partial x} & \dfrac{\partial f_2}{\partial y} \end{vmatrix} \neq 0.$$

This determinant is called the *Jacobian* of the system of simultaneous equations. Our requirement is therefore that the Jacobian may not vanish at any of the x_i, y_i during the iterative process.

If the Jacobian is not zero, then we can solve for h and k as

$$h = \frac{\begin{vmatrix} -f_1(x_i, y_i) & \dfrac{\partial f_1}{\partial y} \\[2mm] -f_2(x_i, y_i) & \dfrac{\partial f_2}{\partial y} \end{vmatrix}}{J} \qquad k = \frac{\begin{vmatrix} \dfrac{\partial f_1}{\partial x} & -f_1(x_i, y_i) \\[2mm] \dfrac{\partial f_2}{\partial x} & -f_2(x_i, y_i) \end{vmatrix}}{J}.$$

Having found h and k, we then use

$$x_{i+1} = x_i + h$$
$$y_{i+1} = y_i + k$$

to get the next approximation, and repeat.

This procedure is the Newton-Raphson method for two simultaneous equations, and is easily expanded for use with three or more equations and variables. When it converges, it has second-order convergence (as did the Newton-Raphson method for one equation) if the roots are simple. But the requirement that the Jacobian is not allowed to be zero is difficult to check without running the program, and so we never know beforehand whether the process will work. Furthermore, the Newton-Raphson method requires many computations and so simpler methods are often used.

4.4 Example 4-1. Newton-Raphson Iteration

As an example, we shall solve the simultaneous equations

$$y = \cos x$$
$$x = \sin y.$$

Put into the standard form, we have

$$f_1(x, y) = \cos x - y = 0$$
$$f_2(x, y) = x - \sin y = 0.$$

Figure 4-3 shows the curves specified by the two equations; the solution is at the intersection of the two curves.

Taking the partial derivatives of each function, we have

$$f_1(x, y) = \cos x - y = 0 \qquad f_2(x, y) = x - \sin y = 0$$

$$\frac{\partial f_1}{\partial x} = -\sin x \qquad\qquad \frac{\partial f_2}{\partial x} = 1$$

$$\frac{\partial f_1}{\partial y} = -1 \qquad\qquad \frac{\partial f_2}{\partial y} = -\cos y,$$

so that the Jacobian is

$$J = \begin{vmatrix} \dfrac{\partial f_1}{\partial x} & \dfrac{\partial f_1}{\partial y} \\[2mm] \dfrac{\partial f_2}{\partial x} & \dfrac{\partial f_2}{\partial y} \end{vmatrix} = \begin{vmatrix} -\sin x & -1 \\[2mm] 1 & -\cos y \end{vmatrix} = \sin x \cos y + 1.$$

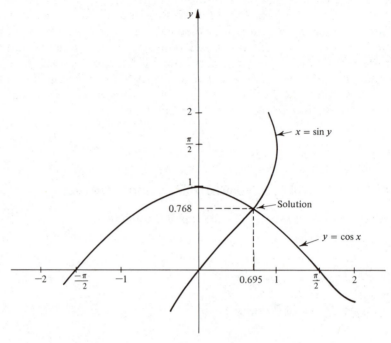

Figure 4-3.   Solution of the simultaneous equations $y = \cos x$ and $x = \sin y$.

In order for the Jacobian to be zero, we must have

$$\sin x \cos y + 1 = 0$$
$$\sin x \cos y = -1,$$

so that either $\sin x$ or $\cos y$ must be negative, with the other being positive; this does not happen in the first quadrant, and so, as long as our values x_i and y_i are either in the first quadrant or on the x and y axes, we can now solve for h and k:

$$h = \frac{\begin{vmatrix} -f_1(x, y) & \dfrac{\partial f_1}{\partial y} \\[2ex] -f_2(x, y) & \dfrac{\partial f_2}{\partial y} \end{vmatrix}}{J} = \frac{\begin{vmatrix} -(\cos x - y) & -1 \\[2ex] -(x - \sin y) & -\cos y \end{vmatrix}}{J}$$

$$= \frac{-(\cos x - y)(-\cos y) + (x - \sin y)(-1)}{\sin x \cos y + 1}$$

$$k = \frac{\begin{vmatrix} \dfrac{\partial f_1}{\partial x} & -f_1(x, y) \\[2mm] \dfrac{\partial f_2}{\partial x} & -f_2(x, y) \end{vmatrix}}{J} = \frac{\begin{vmatrix} -\sin x & -(\cos x - y) \\[2mm] 1 & -(x - \sin y) \end{vmatrix}}{J}$$

$$= \frac{-(x - \sin y)(-\sin x) + (\cos x - y)(1)}{\sin x \cos y + 1}.$$

We now write the following program, starting with the initial guess that $x_0 = y_0 = 1$. Note that a test is included to stop the program if the Jacobian JACOB becomes zero. As we shall see in Chapter 5, even a near-zero value can cause trouble, but for now we shall not worry about that.

```
      REAL   JACOB, K
      X = 1.0
      Y = 1.0
      WRITE (3, 5)
    5 FORMAT (14X, 'X', 14X, 'Y')
      WRITE (3, 10) X, Y
   10 FORMAT (5X, 2F15.8)
      DO 100 I = 1, 5
      JACOB = SIN (X) * COS (Y) + 1.
      IF (JACOB)  15, 200, 15
   15 H = (-(COS (X) - Y) * (-COS (Y))+ (X - SIN (Y)) *(-1.))/ JACOB
      K = (-(X - SIN (Y)) * (-SIN (X))+ (COS (X) - Y) *  1.) / JACOB
      X = X + H
      Y = Y + K
      WRITE (3, 20) I, X, Y
   20 FORMAT (I5, 2F15.8)
  100 CONTINUE
  200 CALL EXIT
      END
```

At each step of the program, the new value of the Jacobian is calculated, tested for zero, and then used to calculate the new values of h and k; these values are then added to the old values of x and y, respectively, to give us new values which are printed.

The program gives us the following results:

	X	Y
	1.00000000	1.00000000
1	0.72027285	0.77568458
2	0.69495215	0.76832706
3	0.69481970	0.76816915
4	0.69481969	0.76816915
5	0.69481969	0.76816915

The Newton-Raphson method provides second-order convergence, and gives us an answer correct to eight decimal places in only four iterations.

4.5 The Modified Newton-Raphson Method

The Newton-Raphson method presented in the last section is not very easy; in general, for n simultaneous equations in n unknowns we must find n^2 partial derivatives, and must then solve for n increments $h, k, \ldots$ by solving n simultaneous equations for n unknowns. We therefore seek an easier method which, while perhaps not converging as quickly, is easier to apply.

The modified Newton-Raphson method we shall now describe consists of applying the single-variable Newton-Raphson method (from Chapter 3) n times, once for each variable. Each time we do this, we assume that the other variables are fixed.

As an example, consider two equations in two unknowns

$$f_1(x, y) = 0$$
$$f_2(x, y) = 0.$$

Taking x_0 and y_0 as initial approximations, we calculate a new value x_1 from the single-variable Newton-Raphson method as

$$x_1 = x_0 - \frac{f_1(x_0, y_0)}{\partial f_1/\partial x},$$

where $\partial f_1/\partial x$ is evaluated at x_0 and y_0. Note that we get a new value x_1 from function f_1 and the most recent values of x and y, in this case from x_0 and y_0.

We now use f_2 and the most recent values of x and y, in this case x_1 and y_0, to calculate a new value y_1:

$$y_1 = y_0 - \frac{f_2(x_1, y_0)}{\partial f_2/\partial y},$$

where $\partial f_2/\partial y$ is evaluated at x_1 and y_0. Now that we have x_1 and y_1, we use f_1 to calculate x_2, and so on.

Note that the choice of using f_1 to calculate a new x and using f_2 to calculate a new y appears arbitrary. Actually, as we shall soon see, it is not arbitrary at all. Generally, we could use either f_1 or f_2 to calculate a new x and use the other function to calculate y. But one of these choices will usually converge while the other diverges, depending on the exact problem. We illustrate this with the following example.

4.6 Example 4-2. Modified Newton-Raphson Method

We now again consider the simultaneous equations from Example 4-1:

$$f_1(x, y) = \cos x - y = 0 \qquad f_2(x, y) = x - \sin y = 0$$

$$\frac{\partial f_1}{\partial x} = -\sin x \qquad\qquad \frac{\partial f_2}{\partial x} = 1$$

$$\frac{\partial f_1}{\partial y} = -1 \qquad\qquad \frac{\partial f_2}{\partial y} = -\cos y.$$

Choosing to use f_1 to calculate a new x and f_2 to calculate a new y, we may write

$$x_{i+1} = x_i - \frac{f_1(x_i, y_i)}{\partial f_1/\partial x} = x_i - \frac{\cos x_i - y_i}{-\sin x_i}$$

$$y_{i+1} = y_i - \frac{f_2(x_{i+1}, y_i)}{\partial f_2/\partial y} = y_i - \frac{x_{i+1} - \sin y_i}{-\cos y_i}.$$

This results in the following program:

```
    X = 1.0
    Y = 1.0
    WRITE (3, 5)
  5 FORMAT (14X, 'X', 14X, 'Y')
    WRITE (3, 10) X, Y
 10 FORMAT (5X, 2F15.8)
    DO 100 I = 1, 10
    X = X - (COS (X) - Y) / (-SIN (X))
    Y = Y - (X - SIN (Y)) / (-COS (Y))
    WRITE (3, 20) I, X, Y
 20 FORMAT (I5, 2F15.8)
100 CONTINUE
    CALL EXIT
    END
```

Unfortunately, this choice leads the iteration to diverge quite badly, as we see from the following results:

	X	Y
	1.00000000	1.00000000
1	0.45369751	0.28230276
2	1.86036233	1.92929063
3	-0.45067813	5.88267911
4	10.98825919	18.23862610
5	29.23468849	54.62950952
6	96.60481718	-231.12493437
7	422.73904311	1720.08429193
8	-1330.71343469	-19633.37488555
9	18927.40093231	****************
10	*****************************	

The solution here is to use f_2 instead of f_1 to calculate new values of x, and to use f_1 instead of f_2 to calculate new values of y:

$$x_{i+1} = x_i - \frac{f_2(x_i, y_i)}{\partial f_2/\partial x} = x_i - \frac{x_i - \sin y_i}{1}$$

$$y_{i+1} = y_i - \frac{f_1(x_{i+1}, y_i)}{\partial f_1/\partial y} = y_i - \frac{\cos x_{i+1} - y_i}{-1},$$

which, incidentally, also is quicker to calculate. The resulting program differs only in the two statements

```
X = X - (X - SIN (Y))
Y = Y - (COS (X) - Y) / (-1)
```

This time the process converges to the same answers as found in Example 4-1:

	X	Y
	1.00000000	1.00000000
1	0.84147098	0.66636674
2	0.61813407	0.81496121
3	0.72769899	0.74670690
4	0.67922554	0.77805945
5	0.70189853	0.76361773
6	0.69153919	0.77026534
7	0.69632570	0.76720406
8	0.69412528	0.76861356
9	0.69513923	0.76796453
10	0.69467251	0.76826337
11	0.69488745	0.76812577
12	0.69478849	0.76818913
13	0.69483405	0.76815995
14	0.69481307	0.76817339
15	0.69482273	0.76816720
16	0.69481828	0.76817005
17	0.69482033	0.76816874
18	0.69481939	0.76816934
19	0.69481982	0.76816907
20	0.69481962	0.76816919
21	0.69481972	0.76816914
22	0.69481967	0.76816916
23	0.69481969	0.76816915
24	0.69481968	0.76816915
25	0.69481969	0.76816915
26	0.69481969	0.76816915

Whereas the complete Newton-Raphson method converged to give 8 place accuracy in only five iterations, the modified Newton-Raphson method does not converge at all if the wrong choice is made as to f_1 and f_2, and converges very slowly (25 iterations to give 8 place accuracy) when the right choice is made.

4.7 Convergence of the Modified Newton-Raphson Method

It is interesting to note why the Newton-Raphson method converges when one choice is made with respect to f_1 and f_2, but diverges when the opposite choice is made.

Figure 4-4 shows the curves defined by $f_1(x, y)$ and $f_2(x, y)$. The point $x_0 = 1.0$ and $y_0 = 1.0$ is the starting point of the iteration, and the modified Newton-Raphson iteration calculates the new approximations in the sequence $x_1, y_1, x_2, y_2, x_3, y_3, \ldots$.

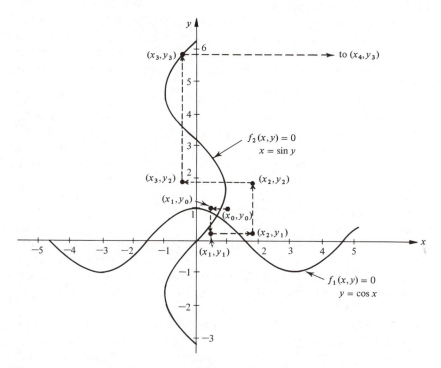

Figure 4-4. Graph of the curves defined by $f_1(x, y) = 0$ and $f_2(x, y) = 0$. The dashed line shows the path taken by the Newton-Raphson iteration as it diverges.

This sequence of values of x_i and y_i can be interpreted as a path in the x–y plane followed by the iteration in its search for the correct intersection of the two curves. Thus the iteration follows the path from (x_0, y_0) through the points

$$(x_1, y_0)$$
$$(x_1, y_1)$$
$$(x_2, y_1)$$
$$(x_2, y_2)$$
$$(x_3, y_2)$$
$$(x_3, y_3)$$
$$\vdots$$

in that sequence; these points are shown in Figure 4-4.

Now, each value of x in the diverging calculation is found from the function $f_1(x, y)$, while each value of y is found from the function $f_2(x, y)$.

Thus each new determination of x starts with a value of y and tries to find a value of x such that the point (x, y) lies on the curve defined by $f_1(x, y) = 0$. But since this curve is only defined for $|y| \leqslant 1$, this cannot always be done. Similarly, each new determination of y starts with the previous value of x and tries to find a value of y such that the point (x, y) lies on the curve defined by $f_2(x, y) = 0$. But sometimes the starting value of x is such that this cannot be done.

This is easily seen by looking at the point (x_1, y_1) in Figure 4-4. This point lies reasonably close to the curve defined by $f_2 = 0$. Now we seek to find a new value x_2 such that (x_2, y_1) will lie on the curve defined by $f_1 = 0$. The one-variable Newton-Raphson iteration which does this happens to be oscillating, and it overshoots the mark, finishing at the point (x_2, y_1). Given this value of x_2, we next try to find a y_2 such that (x_2, y_2) lies on the curve defined by $f_2(x, y) = 0$, but this cannot be done since this curve is only defined between $-1 \leqslant x \leqslant +1$. And so the choice of y_2 must necessarily be invalid, and the process continues.

Reversing f_1 and f_2 solves the problem. Since we now use f_2 to find a value of x given the previous values of x and y, this can be done for any y since f_2 is defined for all values of y. Not only that, the resulting value of x will be less than 1 in magnitude and so will be within range the next time around. Similarly, finding the next y from f_1 can be done for any previous x, and will yield a $|y| \leqslant 1$ which will be satisfactory during the next repetition. Figure 4-5 shows the convergence of the changed method.

This reasoning emphasizes the need to provide a good initial guess x_0, y_0 in the region defined for both functions. But even more important than the question "Why does the iteration diverge when it reaches a region where one of the functions is not defined?" is the question "Why does it reach that region at all?"

Figure 4-6 shows the answer to the latter question. Suppose that either the two functions $f_1 = 0$ and $f_2 = 0$ define two straight lines as shown, or

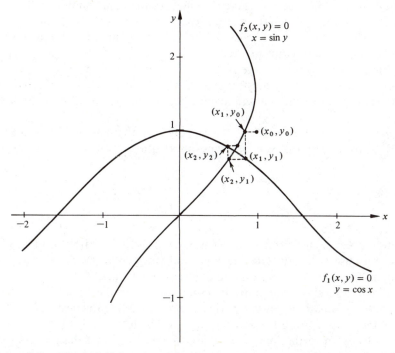

Figure 4-5. The dashed line shows the path taken by the modified Newton-Raphson method when it converges.

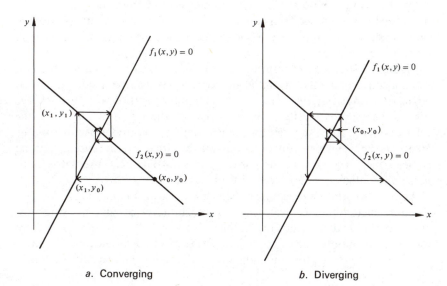

a. Converging   *b.* Diverging

Figure 4-6. Examples of converging and diverging iterations for the same two simultaneous equations.

else that the initial guess is taken so close to the final solution $\bar{x}$, $\bar{y}$ that in that small area both curves may be approximated by straight lines, so that Figure 4-6 holds. In either case, Figure 4-6 represents the ideal case which should have fastest convergence.

But now in Figure 4-6a we pick an initial combination (x_0, y_0) quite far from the actual solution at the intersection of the two curves. We also decide to use the f_1 function to obtain the next x, and the f_2 function to obtain the next y.

Starting from the point (x_0, y_0), we now keep the same y_0 but calculate a new x_1 from f_1 by the Newton-Raphson iteration

$$x_1 = x_0 - \frac{f_1(x_0, y_0)}{\partial f_1/\partial x}.$$

This moves the iteration to the point (x_1, y_0) shown in part a of Figure 4-6. Then we use f_2 to calculate a new y_1, and move up to the f_2 curve at point (x_1, y_1), and so on. This iteration obviously converges.

Figure 4-6b shows the opposite case. This time we pick an (x_0, y_0) very close to the solution, but use f_2 (instead of f_1) to find the next x, and f_1 to find the next y. From the starting point we therefore move left to the line $f_2 = 0$, down to the line $f_1 = 0$, and so on. This time the process quite obviously diverges.

In general, it can easily be shown that the function which has the steeper slope at the solution point $(\bar{x}, \bar{y})$ should be used to find the next x, while the other function should be used to find the next y. If the two functions have the same slope in a neighborhood of the solution, then the iteration will oscillate about a true solution.

Figure 4-7 shows an example of a set of functions for which this method will not converge at all if the initial guess (x_0, y_0) is chosen outside the dashed circle. If the choice of functions is made so that the iteration starts to converge inward, as soon as the iteration path reaches the inside of the circle, the relative slopes of the two functions change, and the initial choice is now wrong. Hence the choice which converges outside the circle diverges inside, and the choice which diverges outside converges inside. The only alternative is to take an initial guess which lies inside the circle.

As you can see, the question of convergence with the modified Newton-Raphson method is a touchy one, and we can never be guaranteed results. On the other hand, provided our initial choice for (x_0, y_0) is reasonably close to the final solution, one of the two choices open to us in picking f_1 and f_2 will usually converge for functions not too peculiar, and so this method is worth a try.

The same remarks apply to use of the modified Newton-Raphson method for more than two equations in two unknowns, but here the question of

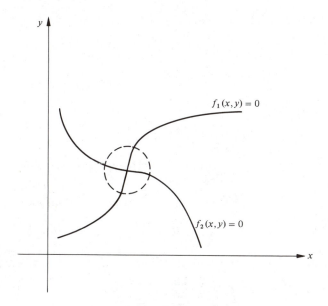

Figure 4-7. One case where the modified Newton-Raphson method may not converge at all.

convergence is even more stringent. In general, for n functions in n unknowns there are $n!$ ways of choosing the n functions to find the n unknowns, and often only one of these choices will converge.

We now give one more two-variable example.

4.8 Example 4-3. Modified Newton-Raphson Method

Figure 4-8 shows the parabola $y = x^2$ and the hyperbola $y = (x + 1)/x$, and we wish to find their intersection. Since both of these equations give y explicitly in terms of x, one way would be by setting

$$x^2 = \frac{x + 1}{x}$$
$$x^3 = x + 1$$
$$x^3 - x - 1 = 0$$

and solving the resulting cubic equation to find its real root.

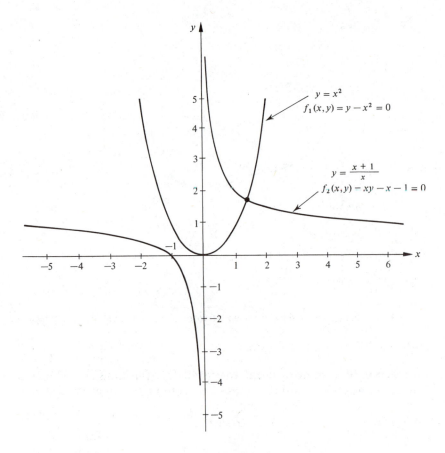

Figure 4-8. The problem of Example 4-3.

We can, however, also attack this problem by solving two simultaneous nonlinear equations

$$y = x^2 \qquad\qquad\qquad y = \frac{x+1}{x}.$$

$$xy = x + 1$$

$$f_1(x, y) = y - x^2 = 0 \qquad f_2(x, y) = xy - x - 1 = 0$$

$$\frac{\partial f_1}{\partial x} = -2x \qquad\qquad \frac{\partial f_2}{\partial x} = y - 1$$

$$\frac{\partial f_1}{\partial y} = 1 \qquad\qquad \frac{\partial f_2}{\partial y} = x$$

We now note that at the solution (Figure 4-8) the curve defined by $f_1(x, y) = 0$ is steeper, and therefore should be used for finding x, while f_2 should be used for finding y. We therefore prepare the iteration

$$x_{i+1} = x_i - \frac{f_1(x_i, y_i)}{\partial f_1/\partial x} \qquad y_{i+1} = y_i - \frac{f_2(x_{i+1}, y_i)}{\partial f_2/\partial y}.$$

As an initial guess, we pick $x_0 = y_0 = 3$ as being a reasonable set of values, and write the following program

```
      X = 3.
      Y = 3.
      WRITE (3, 5)
    5 FORMAT (14X, 'X', 14X, 'Y')
      WRITE (3, 10) X, Y
   10 FORMAT (5X, 2F15.8)
      DO 100 I = 1, 15
      X = X - (Y - X**2) / (-2. * X)
      Y = Y - (X * Y - X - 1.) / X
      WRITE (3, 20) I, X, Y
   20 FORMAT (I5, 2F15.8)
  100 CONTINUE
      CALL EXIT
      END
```

As the following results show, 13 iterations produce an answer accurate to eight decimal places.

	X	Y
	3.00000000	3.00000000
1	2.00000000	1.50000000
2	1.37500000	1.72727272
3	1.31559917	1.76010993
4	1.32673810	1.75372825
5	1.32428632	1.75512370
6	1.32481092	1.75482469
7	1.32469796	1.75488905
8	1.32472225	1.75487521
9	1.32471703	1.75487819
10	1.32471815	1.75487755
11	1.32471791	1.75487769
12	1.32471796	1.75487766
13	1.32471795	1.75487766
14	1.32471795	1.75487766
15	1.32471795	1.75487766

If, on the other hand, we had chosen f_2 to find the next x and f_1 to find the next y, we would have changed only two statements

```
X = X - (X * Y - X - 1.) / (Y - 1.)
Y = Y - (Y - X**2)
```

and found that the program diverges quite badly:

	X	Y
	3.00000000	3.00000000
1	0.50000000	0.25000000
2	-1.33333333	1.77777778
3	1.28571428	1.65306121
4	1.53125001	2.34472660
5	0.74364558	0.55300875
6	-2.23718028	5.00497560
7	0.24968940	0.06234480
8	-1.06649011	1.13740116
9	7.27795873	52.96868340
10	0.01924235	0.00037026

Figure 4-9 shows the paths taken by the converging and diverging iterations; the diverging iteration starts off badly because of the wrong slopes, and soon starts to hit the other hyperbola defined by f_2. The converging iteration, on the other hand, goes directly to the solution.

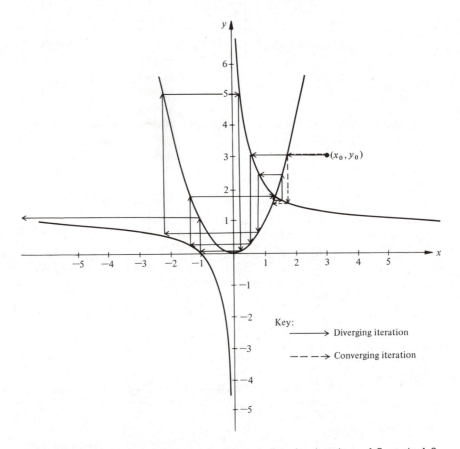

Figure 4-9. Paths taken by the converging and diverging iterations of Example 4-3.

4.9 Complex Roots of Polynomial Equations

In addition to using the techniques of Chapter 3 to find complex roots of a single polynomial, we can also use the method of simultaneous equations. In this section we shall briefly describe an example, without presenting a detailed solution or program.

Example 3-7 found the roots of the polynomial

$$x^2 + x + 1 = 0$$

to be $-0.5 \pm j0.866$. Without knowing this, we may simply assume that the two complex roots are $a + jb$, and its conjugate $a - jb$. The polynomial then consists of the two factors

$$x^2 + x + 1 = 0 = (x - a - jb)(x - a + jb).$$

Now, if $x = a + jb$ then the polynomial should equal zero. Let us substitute this value into the polynomial:

$$
\begin{aligned}
x^2 + x + 1 &= (a + jb)^2 + (a + jb) + 1 = 0 \\
&= a^2 + 2jba + j^2b^2 + a + jb + 1 = 0 \\
&= a^2 + 2jba - b^2 + a + jb + 1 = 0.
\end{aligned}
$$

Collecting real and imaginary terms together,

$$= (a^2 - b^2 + a + 1) + j(2ba + b) = 0.$$

If this equation is to be satisfied, then the real and imaginary parts must both be zero, and we get two simultaneous equations in the two unknowns a and b:

$$
\begin{aligned}
a^2 - b^2 + a + 1 &= 0 = f_1(a, b) \\
2ba + b &= 0 = f_2(a, b).
\end{aligned}
$$

These two simultaneous equations are both nonlinear in a and b, and can be solved using the methods of this chapter to give $a = 0.5$, and $b = \pm 0.866$; since the coefficients are real, we know that the conjugate is also a root, and so both roots are found.

4.10 Problems

1. Try to find solutions to the following simultaneous equations by both the Newton-Raphson method and by the modified Newton-Raphson method. In each case, start by drawing the two curves defined by the equations in the x–y plane, and use the graph to determine good starting values x_0 and y_0. In the case of the modified Newton-Raphson method, use the graph to determine which equation to use for finding the next x_i and which to use for finding y_i.

a. $x^2 + y^2 = 1$
   $x^2 = y$;

b. $x^2 + y^2 = 2$
   $y^2 - x^2 = 1$;

c. $x^2 + y^2 = 1$
   $xy - 1 = 0$;

d. $y - \sin x = 0$
   $y = e^x$;

e. $y = 3x - 2$
   $x = y^{-\frac{1}{4}}$;

f. $3x + 2y - 5 = 0$
   $2x + 3y - 5 = 0$;

g. $y = e^x$
   $y = e^{-2x}$;

h. $y = e^x$
   $x = e^{-y}$;

i. $x = -\tan y$
   $y = e^x - 2\cos x$;

j. $y = x + \sin x$
   $3\sin x - y = 0$.

2. Extend the Newton-Raphson method to three equations in three unknowns, following the same procedure as in this chapter.

5

Matrices, Determinants, and Linear Simultaneous Equations

When we discussed simultaneous equations in Chapter 4, we pointed out that the methods of Chapter 4 were useful for all types of simultaneous equations, linear and nonlinear, but that there were special techniques for handling linear simultaneous equations, which could not be used in the nonlinear case. We will study these now.

5.1 Introduction to Matrices and Determinants

Suppose that we have the following three equations in the three unknowns x_1, x_2, and x_3:

$$3x_1 + 5x_2 - x_3 = 10$$
$$7x_1 - 2x_2 + 3x_3 = 12$$
$$x_1 + 5x_2 - 4x_3 = -1.$$

In this chapter, the notation x_1, x_2, and x_3 has a slightly different meaning from previous chapters. Previously, we would consider x_1 as some particular value of the general variable x, x_2 as some other value of x, and x_3 as some still different value. In this chapter we should think of them instead as three different variables. In fact, we may even be tempted to write the above simultaneous equations as

$$3x + 5y - z = 10$$
$$7x - 2y + 3z = 12$$
$$x + 5y - 4z = -1,$$

which points out that x, y, and z are completely different and really have no connection with each other. There is nothing wrong with this notation, except that as we get to more equations in more unknowns we quickly run out of letters, and the notation becomes a bit unwieldy in some of our methods of solution.

In general, then, we shall call our unknown variables by the symbols x_1, x_2, x_3, and so on. Moreover, we shall identify the constants to the right of the equals sign by the symbols y_1, y_2, and so on. We do this because in many applications they are not constants at all, but may themselves be variables.

For three equations in three unknowns we might therefore write the equations in the form

$$3x_1 + 5x_2 - x_3 = y_1$$
$$7x_1 - 2x_2 + 3x_3 = y_2$$
$$x_1 + 5x_2 - 4x_3 = y_3,$$

where in this particular case $y_1 = 10$, $y_2 = 12$, and $y_3 = -1$.

The coefficients of the variables x are usually written in a form like this:

$$\begin{matrix} 3 & 5 & -1 \\ 7 & -2 & 3 \\ 1 & 5 & -4. \end{matrix}$$

This form is useful in our later calculations, and will be called either a *matrix* or a *determinant*, depending on what we intend to do with it later. In order to make clear which name we intend to use, we will write these coefficients enclosed by brackets when we intend to treat them as a matrix:

$$\begin{bmatrix} 3 & 5 & -1 \\ 7 & -2 & 3 \\ 1 & 5 & -4 \end{bmatrix}.$$

When we intend to treat them as a determinant, we will enclose them between two bars, like this:

$$\begin{vmatrix} 3 & 5 & -1 \\ 7 & -2 & 3 \\ 1 & 5 & -4 \end{vmatrix}.$$

We may often wish to talk of the general case of n equations in n unknowns without wanting to assign specific coefficients to the equations. In this case we will write the equations as

$$a_{11}x_1 + a_{12}x_2 + a_{13}x_3 + \cdots + a_{1n}x_n = y_1$$
$$a_{21}x_1 + a_{22}x_2 + a_{23}x_3 + \cdots + a_{2n}x_n = y_2$$
$$a_{31}x_1 + a_{32}x_2 + a_{33}x_3 + \cdots + a_{3n}x_n = y_3$$
$$\vdots$$
$$a_{n1}x_1 + a_{n2}x_2 + a_{n3}x_3 + \cdots + a_{nn}x_n = y_n.$$

The corresponding coefficient matrix would then also have n^2 coefficients, arranged in n rows and n columns, and would look like this:

$$\begin{bmatrix} a_{11} & a_{12} & a_{13} & \cdots & a_{1n} \\ a_{21} & a_{22} & a_{23} & \cdots & a_{2n} \\ a_{31} & a_{32} & a_{33} & \cdots & a_{3n} \\ \vdots & & & & \vdots \\ a_{n1} & a_{n2} & a_{n3} & \cdots & a_{nn} \end{bmatrix} = [A].$$

Each of the terms in the matrix is a different number, and has a subscript which indicates its position in the matrix (or determinant). We can talk of a particular term as the term a_{ij}, where i is the number of the row and j is the number of the column where that term is located.

For ease in notation, we may also call an entire matrix or determinant by a capital letter name. For example, a matrix of coefficients a_{ij} might be called the matrix $[A]$; a determinant of coefficients which we call c_{ij} might be called the determinant $|C|$. Note that the same letter is generally used, except that a lower-case letter with subscripts is used for the individual elements, and a capital letter is used for the overall name. Moreover, brackets are again used for the matrix and two bars for the determinant.

5.2 Properties of Matrices

Although matrices and determinants may look alike, they are used for different applications, and therefore have somewhat different properties. From now on we shall treat them separately.

1. Matrices may have any number of rows and any number of columns, and the number of rows need not be the same as the number of columns. If a matrix has m rows and n columns, then we say it is an $m \times n$ matrix (pronounced "m by n"). If the number of rows is the same as the number of columns, we say that the matrix is *square*; otherwise it is *rectangular*. An $m \times 1$ matrix, which has only one column, is called a *column matrix*; a $1 \times n$ matrix, which has only one row, is called a *row matrix*.
2. Two matrices are equal only if (a) they are the same size (same number of rows and columns), and (b) corresponding terms in the two matrices are equal.
3. A zero matrix is one all of whose terms are 0.
4. The sum of two matrices (defined only when the two matrices are the same size) is a new matrix each of whose elements is the sum of the two corresponding elements in the original matrices.

   EXAMPLE:

   $$\begin{bmatrix} 1 & 2 & 3 \\ 2 & 3 & 4 \\ 3 & 4 & 5 \end{bmatrix} + \begin{bmatrix} 1 & 1 & 1 \\ 2 & 2 & 2 \\ 3 & 3 & 3 \end{bmatrix} = \begin{bmatrix} 2 & 3 & 4 \\ 4 & 5 & 6 \\ 6 & 7 & 8 \end{bmatrix}.$$

5. Since it would be convenient if we could say that $2[A] = [A] + [A]$, we define the product of a constant times a matrix as a new matrix where each term is obtained by multiplying the corresponding term of the original matrix by the constant.

   EXAMPLE:

   $$2\begin{bmatrix} 1 & 2 & 3 \\ 2 & 3 & 4 \\ 3 & 4 & 5 \end{bmatrix} = \begin{bmatrix} 2 & 4 & 6 \\ 4 & 6 & 8 \\ 6 & 8 & 10 \end{bmatrix}.$$

6. The difference of two matrices (defined only when the two matrices are the same size) is a new matrix each of whose terms is the difference between the corresponding terms in the original matrices.

EXAMPLE:

$$\begin{bmatrix} 1 & 2 & 3 \\ 2 & 3 & 4 \\ 3 & 4 & 5 \end{bmatrix} - \begin{bmatrix} 1 & 1 & 1 \\ 2 & 2 & 2 \\ 3 & 3 & 3 \end{bmatrix} = \begin{bmatrix} 0 & 1 & 2 \\ 0 & 1 & 2 \\ 0 & 1 & 2 \end{bmatrix}.$$

7. The product of two matrices is defined only when the number of columns in the first matrix is equal to the number of rows in the second. If an $m \times n$ matrix $[A]$ is multiplied by an $n \times p$ matrix $[B]$, the product matrix $[C]$ is an $m \times p$ matrix where each term is defined by

$$c_{ij} = \sum_{k=1}^{n} a_{ik} b_{kj}.$$

EXAMPLE:

$$\begin{bmatrix} 1 & 2 & 0 \\ 3 & -4 & 1 \end{bmatrix} \begin{bmatrix} 1 & 2 \\ -1 & 1 \\ 0 & 1 \end{bmatrix} =$$

$$\begin{bmatrix} 1-2+0 & 2+2+0 \\ 3+4+0 & 6-4+1 \end{bmatrix} = \begin{bmatrix} -1 & 4 \\ 7 & 3 \end{bmatrix}.$$

In other words, we might say that the ith row, jth column term of $[C]$ is obtained by multiplying the ith row of $[A]$ by the jth column of $[B]$, term-by-term, and adding the products.

Note that even if $[A][B]$ is defined, the product $[B][A]$ may not be defined (if $[A]$ and $[B]$ do not have the right numbers of rows and columns). Moreover, a simple multiplication of two square matrices of the same size will show that even if $[B][A]$ is defined, it need not equal $[A][B]$.

8. We define an *identity* or *unit matrix* $[I]$ such that it is a square matrix where all terms are 0 except those on the *main diagonal*, which are 1. The main diagonal consists of those terms I_{ij} for which $i = j$.

EXAMPLES:

$$\begin{bmatrix} 1 & 0 \\ 0 & 1 \end{bmatrix} \quad \begin{bmatrix} 1 & 0 & 0 \\ 0 & 1 & 0 \\ 0 & 0 & 1 \end{bmatrix} \quad \begin{bmatrix} 1 & 0 & 0 & 0 \\ 0 & 1 & 0 & 0 \\ 0 & 0 & 1 & 0 \\ 0 & 0 & 0 & 1 \end{bmatrix}.$$

The reason for this definition is that it makes

$$[I][A] = [A][I] = [A],$$

which is what we would expect for a unit matrix. (But note: for this multiplication to hold, we must pick a unit matrix of the right size to permit a valid multiplication.)

9. The *transpose* of a matrix $[A]$ is the new matrix, called $[A]^T$, where the rows and columns of $[A]$ are interchanged.

EXAMPLE:

$$[A] = \begin{bmatrix} 1 & 2 & 3 \\ 4 & 5 & 6 \\ 7 & 8 & 9 \end{bmatrix}; \quad [A]^T = \begin{bmatrix} 1 & 4 & 7 \\ 2 & 5 & 8 \\ 3 & 6 & 9 \end{bmatrix}.$$

5.3 Example 5-1. Simple Matrix Manipulations in FORTRAN

Since the general matrix is two-dimensional, it can easily be represented in FORTRAN as a two-dimensional array. To define a 3×4 matrix $[A]$, for example, we use the FORTRAN statement DIMENSION A(3, 4) to define an array A. Then we can refer to the element a_{ij} of the array using the form A(I, J). The following FORTRAN IV program shows how to do simple matrix addition, subtraction, and multiplication, using the two 3×3 matrices

$$[A] = \begin{bmatrix} 1 & 1 & 1 \\ 2 & 2 & 1 \\ 3 & 2 & 1 \end{bmatrix} \quad [B] = \begin{bmatrix} 0 & -1 & 1 \\ -1 & 2 & -1 \\ 2 & -1 & 0 \end{bmatrix}.$$

```
C       MATRIX ADDITION, SUBTRACTION, AND MULTIPLICATION
        DIMENSION A(3,3), B(3,3), C(3,3)
C       DEFINE MATRIX ELEMENTS
        A(1, 1) = 1
        A(1, 2) = 1
        A(1, 3) = 1
        A(2, 1) = 2
        A(2, 2) = 2
        A(2, 3) = 1
        A(3, 1) = 3
        A(3, 2) = 2
        A(3, 3) = 1
        B(1, 1) = 0
        B(1, 2) = -1
```

```
                B(1, 3) = 1
                B(2, 1) = -1
                B(2, 2) = 2
                B(2, 3) = -1
                B(3, 1) = 2
                B(3, 2) = -1
                B(3, 3) = 0
      C         CALCULATE AND PRINT SUM
                DO 10 IROW = 1, 3
                DO 10 JCOL = 1, 3
      10 C(IROW, JCOL) = A(IROW, JCOL) + B(IROW, JCOL)
                WRITE (3, 14)
      14 FORMAT (' SUM')
                WRITE (3, 15) ((C(I, J), J = 1, 3), I = 1, 3)
      15 FORMAT (3(3F6.1, /))
      C         CALCULATE AND PRINT OUT DIFFERENCE
                DO 20 IROW = 1, 3
                DO 20 JCOL = 1, 3
      20 C(IROW, JCOL) = A(IROW, JCOL) - B(IROW, JCOL)
                WRITE (3, 25)
      25 FORMAT (' DIFFERENCE')
                WRITE (3, 15) ((C(I, J), J = 1, 3), I = 1, 3)
      C         CALCULATE AND PRINT OUT PRODUCT A * B
                DO 50 IROW = 1, 3
                DO 50 JCOL = 1, 3
                SUM = 0
                DO 40 K = 1, 3
      40 SUM = SUM + A(IROW, K) * B(K, JCOL)
      50 C(IROW, JCOL) = SUM
                WRITE (3, 55)
      55 FORMAT (' PRODUCT A * B')
                WRITE (3, 15) ((C(I, J), J = 1, 3), I = 1, 3)
                CALL EXIT
                END
```

The sum, difference, and product matrices provided by the computer are as follows:

```
SUM
   1.0    0.0    2.0
   1.0    4.0    0.0
   5.0    1.0    1.0

DIFFERENCE
   1.0    2.0    0.0
   3.0    0.0    2.0
   1.0    3.0    1.0

PRODUCT A * B
   1.0    0.0    0.0
   0.0    1.0    0.0
   0.0    0.0    1.0
```

An interesting point is that the product of $[A]$ and $[B]$ gives us the unit matrix $[I]$. If this happened to us with two real numbers, we would say that they are reciprocals of each other; in matrix terminology, we say that $[A]$ and $[B]$ are *inverses* of each other if their product is the identity matrix $[I]$. To find the inverse of a matrix, we need a knowledge of determinants.

5.4 Properties of Determinants

Although a determinant may look like a matrix, its properties are completely different. For one, a determinant is actually a single number, although it may be written down in a form similar to a matrix.

1. When written in a matrix-like form, a determinant always has the same number of rows as columns; thus it is always a square array of numbers. An $n \times n$ determinant is said to be of nth order.
2. Whereas the matrix is merely an ordered way of writing a collection of numbers, a determinant actually represents a single number, which can be calculated in a specified way from the array of numbers written in a matrix-like form. While we may often loosely use the term *determinant* when referring to something like

$$\begin{vmatrix} 1 & 2 & 3 \\ 3 & 3 & 2 \\ 1 & 3 & 7 \end{vmatrix}$$

we are subconsciously thinking of the determinant's numerical value instead.

Most high-school students know how to evaluate a second- or third-order determinant; nevertheless, the definition of the value of a determinant is of interest since it shows how impractical it is in computer evaluation.

In evaluating an nth-order determinant, consider first the n integers 1, 2, 3, ..., n. In their normal order, each integer is larger than any preceding. Any rearrangement of these integers out of their normal order is called a *permutation*, and it results in the fact that some of the smaller integers appear after the larger integers.

With n integers, there are $n!$ possible permutations, since there are $n!$ possible different ways of rearranging these integers. Let us call each instance when a smaller integer lies to the right of a larger integer an *inversion*. Thus, for example, the permutation 1342 has two inversions since the digits 3–2 and 4–2 are the only two pairs for which the smaller integer lies to the right of the larger one. Any particular permutation of n integers has a specific number k of inversions.

We can now define the value of a determinant as the sum

$$|A| = \sum^{n!} (-1)^k a_{1\alpha} a_{2\beta} a_{3\gamma} \cdots a_{n\omega},$$

where the sequence $\alpha, \beta, \gamma, \ldots, \omega$ represents one of the $n!$ permutations of the n integers, k is the number of inversions in that permutation, and the a's are the terms in the determinant.

As an example, the second-order determinant

$$|A| = \begin{vmatrix} a_{11} & a_{12} \\ a_{21} & a_{22} \end{vmatrix}$$

$$= (-1)^0 a_{11}a_{22} + (-1)^1 a_{12}a_{21}$$

$$= a_{11}a_{22} - a_{12}a_{21},$$

which is a result we already know. Further, the third-order determinant

$$|A| = \begin{vmatrix} a_{11} & a_{12} & a_{13} \\ a_{21} & a_{22} & a_{23} \\ a_{31} & a_{32} & a_{33} \end{vmatrix}$$

$$= (-1)^0 a_{11}a_{22}a_{33} + (-1)^1 a_{11}a_{23}a_{32} + (-1)^2 a_{13}a_{21}a_{32}$$
$$+ (-1)^3 a_{13}a_{22}a_{31} + (-1)^4 a_{12}a_{23}a_{31} + (-1)^5 a_{12}a_{21}a_{33}$$

$$= a_{11}a_{22}a_{33} - a_{11}a_{23}a_{32} + a_{13}a_{21}a_{32} - a_{13}a_{22}a_{31}$$
$$+ a_{12}a_{23}a_{31} - a_{12}a_{21}a_{33},$$

where the number of inversions comes from the fact that the sequences 123, 132, 312, 321, 231, and 213 are each obtained from the preceding permutation by one additional inversion.

In the case of the third-order determinant, there is a simple geometric way of obtaining the six terms in the summation from the determinant by re-writing the determinant as follows, and then taking products along the diagonals as shown.

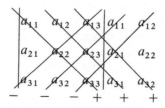

$$-\quad -\quad -\quad +\quad +\quad +$$

But this geometric method no longer works for higher orders, and so we must return to the series definition. Unfortunately, while rather mathematically elegant, this definition of a determinant is useless when applied to a computer solution. Since there are $n!$ permutations, there are $n!$ terms in the summation, and each term requires $(n - 1)$ multiplications, for a total of $n!(n - 1)$ multiplications, which are usually performed in real (floating-point) arithmetic, and thus can be quite slow. Added to this is the problem of generating $n!$ permutations and calculating the number of inversions for each one. Moreover, there are another $(n)(n!)$ integer multiplications and additions not so immediately obvious: Whenever we specify a particular element $A(I, J)$ in a two-dimensional FORTRAN array, the computer performs an integer multiplication and addition to find that element in the computer memory. Hence even a modest 10×10 determinant, for which $10! = 3,628,800$, can use a large amount of computer time!

We shall return to the problem of determinant evaluation after some more properties.

3. Two determinants are equal if their values are equal. Thus, unlike a matrix, the two determinants need not resemble each other in any other way. For example,

$$\begin{vmatrix} 1 & 0 & 0 \\ 0 & 1 & 0 \\ 0 & 0 & 1 \end{vmatrix} = \begin{vmatrix} 2 & 1 \\ 1 & 1 \end{vmatrix} = 1.$$

4. A zero determinant need not have any zero elements, as in

$$\begin{vmatrix} 1 & 1 \\ 1 & 1 \end{vmatrix};$$

on the other hand, if any one of the rows or columns of the determinant is zero, then its value is zero. This is so since each of the terms in the summation which defines the determinant (property 2) will have a zero factor, and will therefore be zero.

5. Multiplying any row, or column, of a determinant by a constant multiplies the value of the entire determinant by the same constant. This can be shown to be true since each term in the defining summation will contain that constant as an additional factor.

6. The sum, difference, product, or quotient of two determinants is defined in terms of the sum, difference, product or quotient of their values.

7. The product of two determinants may also be defined the same way as the product of two matrices.

8. Adding any row (or multiple of a row) to any other row, or adding any column (or multiple thereof) to any other column, does not change the value of the determinant. This is a consequence of the $(-1)^k$ term in the defining summation and can be proven from the summation with a little effort.

9. Interchanging two rows or two columns of a determinant does not change the absolute value of the determinant, but does change the sign.

10. If two rows (or columns) are either equal or one is a multiple of the other, then the determinant is equal to zero, since we could use property 8 to make one of the rows (or columns) equal to zero.

5.5 Evaluation of Determinants

The defining summation presented in the last section for evaluating the determinant is moderately useful for hand evaluation of determinants up to perhaps fourth or fifth order, except that checking that all permutations have been used and finding their number of inversions is somewhat clumsy. Another method, called *expansion by minors*, is exactly identical mathematically, but provides a more systematic way of attacking the problem.

Let us start with an nth-order determinant $|A|$, and delete the ith row and jth column. This results in an $(n-1)$th-order determinant which we shall call the *minor* $|M_{ij}|$.

Furthermore, let

$$|A_{ij}| = (-1)^{i+j}|M_{ij}|$$

be called the *signed minor*, or *cofactor* of the term a_{ij} which appears in the row and column crossed out.

By proper manipulation of the defining summation for a determinant, we could write the value of the determinant as

$$|A| = a_{i1}|A_{i1}| + a_{i2}|A_{i2}| + a_{i3}|A_{i3}| + \cdots + a_{in}|A_{in}|.$$

This is called an expansion by minors along the ith row. Thus we expand along the ith row by multiplying the first term by its signed minor, adding the product of the second term by *its* signed minor, and so on. We could also expand along a column as

$$|A| = a_{1j}|A_{1j}| + a_{2j}|A_{2j}| + a_{3j}|A_{3j}| + \cdots + a_{nj}|A_{nj}|.$$

Expansion by minors therefore lets us write an nth-order determinant as the sum of n determinants, each of order $(n-1)$. In turn, each of these determinants can be expanded in terms of order $(n-2)$ determinants, and so on.

We may expand a determinant along any row or column. If any row or column has a large number of zero terms, expansion along that row or column is particularly easy. For example, in the case of the fourth-order determinant.

$$\begin{vmatrix} 1 & 0 & 3 & 2 \\ 0 & 1 & 2 & 1 \\ 4 & 0 & 3 & 3 \\ 1 & 4 & 2 & 2 \end{vmatrix}$$

we can expand along the second column to get

$$\underset{a_{22}}{\underset{\uparrow}{1}}\underset{|A_{22}|}{\underset{\uparrow}{\begin{vmatrix} 1 & 3 & 2 \\ 4 & 3 & 3 \\ 1 & 2 & 2 \end{vmatrix}}} + \underset{a_{42}}{\underset{\uparrow}{4}}\underset{|A_{42}|}{\underset{\uparrow}{\begin{vmatrix} 1 & 3 & 2 \\ 0 & 2 & 1 \\ 4 & 3 & 3 \end{vmatrix}}}$$

Each of the third-order signed minors could in turn be expanded in terms of second-order determinants.

Expansion by minors is an easy-to-remember, systematic method. Nevertheless, if we were to apply it to a general nth-order determinant and write out all the intermediate steps involved in breaking up that determinant into determinants of order $(n - 1)$, $(n - 2)$, $(n - 3)$, ... 2, and 1, we would in effect be performing the summation by which the determinant is defined. Hence expansion by minors is just as impractical for large matrices as use of the defining summation.

Yet, the method provides a useful clue toward a practical method. But first a few definitions:

Definition *A matrix or determinant, all of whose terms either above or below the main diagonal are zero, is called a* triangular *matrix or determinant.*

Definition *An* upper triangular *matrix or determinant has zero terms below the main diagonal.*

Definition *A* lower triangular *matrix or determinant has zero terms above the main diagonal.*

For example, the two determinants below are upper and lower triangular determinants, respectively.

$$
\begin{vmatrix} 1 & 2 & 3 & 4 \\ 0 & 4 & 3 & 2 \\ 0 & 0 & 5 & 8 \\ 0 & 0 & 0 & -1 \end{vmatrix}
\qquad
\begin{vmatrix} 5 & 0 & 0 & 0 & 0 \\ 3 & 10 & 0 & 0 & 0 \\ 0 & 4 & 2 & 0 & 0 \\ 1 & 2 & 3 & 4 & 0 \\ 5 & 5 & 5 & 5 & 5 \end{vmatrix}
$$

Triangular determinants are important because for them the expansion by minors is particularly easy due to the large number of zero terms. In fact, for a triangular determinant $|A|$, we have

$$|A| = a_{11}a_{22}a_{33}a_{44} \cdots a_{nn}.$$

As a demonstration, consider an upper triangular determinant $|A|$. Expanding $|A|$ along the first column, we have

$$|A| = a_{11}|A_{11}| + 0 + 0 + 0 + 0 + \cdots,$$

since all the other terms in the first column are zero. The cofactor $|A_{11}|$ is itself triangular and can be expanded along its first column to give

$$|A_{11}| = a_{22}|B_{22}| + 0 + 0 + 0 + \cdots,$$

where $|B_{22}|$ is an $(n - 2)$-order determinant obtained from $|A|$ by deleting the first two rows and columns. Since each succeeding cofactor is itself triangular, we eventually reach the conclusion that the value of the determinant is merely the product of the terms along the main diagonal.

For example:

$$
\begin{vmatrix} 5 & 3 & 2 & 2 \\ 0 & -1 & 3 & 3 \\ 0 & 0 & 2 & 5 \\ 0 & 0 & 0 & 6 \end{vmatrix}
= 5 \begin{vmatrix} -1 & 3 & 3 \\ 0 & 2 & 5 \\ 0 & 0 & 6 \end{vmatrix}
$$

$$
= 5 \left\{ -1 \begin{vmatrix} 2 & 5 \\ 0 & 6 \end{vmatrix} \right\}
$$

$$
= 5\{-1[(2)\ (6)]\}
$$

$$
= (5)(-1)(2)(6)
$$

$$
= -60.
$$

This general result can be applied to any determinant if we note that *any determinant can be made into a triangular determinant by some simple manipulations*. These are based on two properties of determinants discussed in the previous section:

Property 7 *Adding any row (or multiple of a row) to any other row, or adding any column (or multiple thereof) to any other column, does not change the value of a determinant.*

Property 8 *Interchanging two rows or two columns of a determinant does not change the absolute value of the determinant, but changes the sign.*

EXAMPLE: Consider the third-order determinant

$$\begin{vmatrix} 1 & 1 & 1 \\ 2 & 2 & 1 \\ 3 & 2 & 1 \end{vmatrix}.$$

Without changing the determinant, we can multiply the first row by -2 and add the result to the second row; this is equivalent to subtracting twice the first row from the second. We can also subtract three times the first row from the third:

$$\begin{vmatrix} 1 & 1 & 1 \\ 2 & 2 & 1 \\ 3 & 2 & 1 \end{vmatrix} = \begin{vmatrix} 1 & 1 & 1 \\ 0 & 0 & -1 \\ 0 & -1 & -2 \end{vmatrix}.$$

Finally, interchanging the second and third rows yields (notice the minus sign)

$$\begin{vmatrix} 1 & 1 & 1 \\ 2 & 2 & 1 \\ 3 & 2 & 1 \end{vmatrix} = - \begin{vmatrix} 1 & 1 & 1 \\ 0 & -1 & -2 \\ 0 & 0 & -1 \end{vmatrix}.$$

In the resulting triangular matrix, we multiply out the terms on the main diagonal to get $+1$, so that the determinant equals -1.

This method of evaluating a determinant is generally called a *Gaussian elimination*, and can be expressed in a more general way as follows:

Suppose we have the general determinant

$$|A| = \begin{vmatrix} a_{11} & a_{12} & a_{13} & \cdots & a_{1n} \\ a_{21} & a_{22} & a_{23} & \cdots & a_{2n} \\ a_{31} & a_{32} & a_{33} & \cdots & a_{3n} \\ a_{41} & a_{42} & a_{43} & \cdots & a_{4n} \\ \vdots & & & & \vdots \\ a_{n1} & a_{n2} & a_{n3} & \cdots & a_{nn} \end{vmatrix}$$

to be evaluated. In the first step we wish to make all terms in the first column *except* a_{11} equal to zero. Hence a_{11} itself is not allowed to be zero; thus, before we start, we look at a_{11} and, if it is zero, we interchange the first row with some other row whose first term is nonzero. In fact, we generally interchange with that row whose first term is the largest in magnitude, regardless of whether a_{11} was zero or not, since this results in less roundoff error in succeeding calculations. Of course, switching two rows will change the sign, which we must remember.

Having done this, we now multiply the first row by the constant a_{21}/a_{11} and subtract the result from the second row. Hence the second row becomes

$$\left(a_{21} - \frac{a_{21}}{a_{11}}a_{11}\right)\left(a_{22} - \frac{a_{21}}{a_{11}}a_{12}\right)\left(a_{23} - \frac{a_{21}}{a_{11}}a_{13}\right)\cdots\left(a_{2n} - \frac{a_{21}}{a_{11}}a_{1n}\right),$$

where the first term is therefore zero.

In the same way, we multiply the first row by the constant a_{31}/a_{11} and subtract the result from the third row. As before, the a_{31} term now becomes zero. Repeating for each row of the determinant, we finally obtain the new determinant

$$|A| = \begin{vmatrix} a_{11} & a_{12} & a_{13} & \cdots & a_{1n} \\ 0 & a'_{22} & a'_{23} & \cdots & a'_{2n} \\ 0 & a'_{32} & a'_{33} & \cdots & a'_{3n} \\ 0 & a'_{42} & a'_{43} & \cdots & a'_{4n} \\ \vdots & & & & \vdots \\ 0 & a'_{n2} & a'_{n3} & \cdots & a'_{nn} \end{vmatrix}$$

where the primes indicate that the first row has stayed the same, but every other row has been changed.

In this procedure, the a_{11} term plays an important role in making the rest of the first column zero, since it is a multiple of a_{11} that is subtracted from the other rows. For this reason, a_{11} is called a *pivot* and the above reduction is called a *pivotal reduction* or *pivotal condensation*.

The next step is to make all except the top two terms of the second column equal to zero. Leaving the first row unchanged, we examine the other rows to find the largest magnitude term in the second column and, if necessary, interchange two rows to bring that term into the second row as the new a'_{22}. In essence, we are repeating the previous procedure, except that a'_{22} is the new pivot.

We now multiply the second row by a'_{32}/a'_{22} and subtract the result from the third row, multiply the second row by a'_{42}/a'_{22} and subtract the result from the fourth row, and so on, until we obtain the determinant

$$|A| = \begin{vmatrix} a_{11} & a_{12} & a_{13} & \cdots & a_{1n} \\ 0 & a'_{22} & a'_{23} & \cdots & a'_{2n} \\ 0 & 0 & a''_{33} & \cdots & a''_{3n} \\ 0 & 0 & a''_{43} & \cdots & a''_{4n} \\ \vdots & & & & \vdots \\ 0 & 0 & a''_{n3} & \cdots & a''_{nn} \end{vmatrix}$$

where the double primes from the third row down again indicate that the terms have changed.

We proceed in this way, next using a''_{33} as the pivot, then using a'''_{44} and so on, until we run out of rows. As the last step, we take the product of all the terms on the main diagonal to evaluate the determinant:

$$A = a_{11}a'_{22}a''_{33}a'''_{44} \cdots a_{nn}^{(n-1)}(-1)^s.$$

We also note that if s interchanges of two rows are performed, then the sign is modified by the factor $(-1)^s$.

5.6 Example 5-2. Evaluating a Determinant by Gaussian Elimination

As might be expected, the program to evaluate a determinant using the Gaussian elimination method (pivotal reduction) is somewhat involved and lengthy. As an example, we shall evaluate the fourth-order determinant

$$\begin{vmatrix} -1 & 3 & 5 & 2 \\ 1 & 9 & 8 & 4 \\ 0 & 1 & 0 & 1 \\ 2 & 1 & 1 & -1 \end{vmatrix} = 14.$$

The program which follows is fairly general and could be extended for larger determinants as well, simply by changing the first four statements to accept larger determinants and read more data, and either changing the 320 FORMAT statement, or else deleting it and its corresponding WRITE statement altogether.

The program essentially consists of the following:

1. Reading data and initializing. One of the constants to be initialized is SIGN, which starts at $+1$. Each time a row interchange is done later, SIGN is changed to $-$SIGN; in the end, SIGN is multiplied into the value of the determinant.
2. A loop repeated $(n-1)$ times, once for each row to be reduced, performs the following:
   a. Checks for the largest pivot;
   b. Exits if no nonzero pivot can be found;
   c. Interchanges two rows, if necessary to maximize the pivot;
   d. Performs a pivotal reduction on the remaining rows.
3. Computation of the determinant and printout of the final triangular determinant and its value.

If at any time the program cannot find a nonzero pivot for the next row reduction, that indicates that all the remaining elements in the column considered are already zero, and that no reduction need actually be done. But since all possible pivotal elements are zero, we can easily show that the determinant must be equal to zero, and so the program exits.

The overall program is as follows:

```
            DIMENSION A(4, 4)
            N = 4
      C     READ TERMS FROM CARD
            READ (2, 10) ((A(I, J), J = 1, 4), I = 1, 4)
         10 FORMAT (16F5.0)
            SIGN = 1.
            LAST = N - 1
      C     START OVERALL LOOP FOR (N-1) PIVOTS
            DO 200 I = 1, LAST
      C
      C     FIND THE LARGEST REMAINING TERM IN I-TH COLUMN FOR PIVOT
      C
            BIG = 0.
            DO 50 K = I, N
            TERM = ABS (A(K, I))
            IF (TERM - BIG) 50, 50, 30
         30 BIG = TERM
            L = K
         50 CONTINUE
      C     CHECK WHETHER A NON-ZERO TERM HAS BEEN FOUND
            IF (BIG) 80, 60, 80
         60 CALL EXIT
      C     L-TH ROW HAS THE BIGGEST TERM -- IS I = L
         80 IF (I - L) 90, 120, 90
      C     I IS NOT EQUAL TO L, SWITCH ROWS I AND L
```

```
   90 SIGN = - SIGN
      DO 100 J = 1, N
      TEMP = A(I, J)
      A(I, J) = A(L, J)
  100 A(L, J) = TEMP
C
C        NOW START PIVOTAL REDUCTION
C
  120 PIVOT = A(I, I)
      NEXTR = I + 1
C        FOR EACH OF THE ROWS AFTER THE I-TH
      DO 200 J = NEXTR, N
C        MULTIPLYING CONSTANT FOR THE J-TH ROW IS
      CONST = A(J, I) / PIVOT
C        NOW REDUCE EACH TERM OF THE J-TH ROW
      DO 200 K = I, N
  200 A(J, K) = A(J, K) - CONST * A(I, K)
C        END OF PIVOTAL REDUCTION - NOW COMPUTE DETERMINANT
      DET = SIGN
      DO 300 I = 1, N
  300 DET = DET * A(I , I)
C        PRINT THE RESULTING TRIANGULAR DETERMINANT
      WRITE (3, 320) ((A(I, J), J = 1, N), I = 1, N)
  320 FORMAT (4F12.6, /)
C        PRINT THE VALUE OF THE DETERMINANT
      WRITE (3, 350) DET
  350 FORMAT (F15.6)
      CALL EXIT
      END
```

For the determinant

$$\begin{vmatrix} -1 & 3 & 5 & 2 \\ 1 & 9 & 8 & 4 \\ 0 & 1 & 0 & 1 \\ 2 & 1 & 1 & -1 \end{vmatrix} = 14$$

we obtain the following reduced triangular determinant, followed by the value, which comes out as 14.000001 instead of exactly 14.0. Using standard precision arithmetic, the error is quite small in this particular instance.

```
 2.000000    1.000000    1.000000   -1.000000

 0.000000    8.500001    7.500000    4.500000

 0.000000    0.000000    2.411765   -0.352941

 0.000000    0.000000   -0.000000    0.341463

     14.000001
```

Since all the terms in this particular determinant were integers, we are tempted to perform the entire program in integer arithmetic, but of course this will not work, as we can see from the reduced determinant.

Even on such a small determinant, we already begin to suspect that roundoff error might be troublesome. We note, for example, that the a_{43}''' term somehow appears to be -0.000000 instead of just 0.0. Apparently if we had printed more than six decimal places we would have seen a tiny error here, due entirely to roundoff error. Quite obviously, a larger determinant would have had even more roundoff error.

The very simple ploy of looking for the largest pivot term instead of using just any pivot that happens to come along, for example, has made a difference. For example, when we rerun the program without the portion which sorts the possible pivots to find the largest, we obtain the following results:

```
-1.000000    3.000000    5.000000    2.000000

 0.000000   12.000001   13.000001    6.000000

 0.000000    0.000000   -1.083333    0.500000

 0.000000    0.000000    0.000000    1.076924

        14.000009
```

The results show an error about 9 times larger. But what is not so obvious is the fact that, under some conditions, the roundoff error can be so bad that the results are either wrong, or absolutely misleading and useless. This condition is rather dangerous, and we will examine it closer in a few pages.

5.7 Cramer's Rule for Solving Simultaneous Equations

The most well-known rule for solving simultaneous linear equations is Cramer's rule, which we mention here just for the sake of completeness; actually it is not too useful for large systems of equations.

Suppose we have the two simultaneous equations

$$ax + by = e$$
$$cx + dy = f,$$

where x and y are the unknowns, and a, b, c, d, e, and f are known quantities (constants). The simplest way of solving for one variable, say x, is to multiply the top equation by d, multiply the bottom equation by b, and subtract the bottom equation from the top:

$$
\begin{aligned}
adx + bdy &= de \\
- (bcx + bdy &= bf) \\
\hline
adx - bcx &= de - bf \\
(ad - bc)x &= de - bf \\
x &= \frac{de - bf}{ad - bc}.
\end{aligned}
$$

We could write the numerator and denominator each as a second-order determinant

$$x = \frac{\begin{vmatrix} e & b \\ f & d \end{vmatrix}}{\begin{vmatrix} a & b \\ c & d \end{vmatrix}}.$$

This is the way we usually see Cramer's rule applied. The denominator is simply the matrix of the coefficients of the equations, while the numerator is the same matrix of numbers, except that the column corresponding to the coefficients of x is replaced by the column of constants from the right of the equals sign.

It now appears that the solution for x, or the solution for y, which is

$$y = \frac{\begin{vmatrix} a & e \\ c & f \end{vmatrix}}{\begin{vmatrix} a & b \\ c & d \end{vmatrix}}$$

is valid only when the denominator is nonzero, so that the division is valid. But this is not quite true. Let us consider two separate cases:

1. The quantities e and f are not both zero. We call the resulting system of equations *nonhomogeneous*.
2. Both e and f are zero. We call the resulting system *homogeneous*.

Now, in the nonhomogeneous case, which is the more common, suppose that we find x as the quotient of two determinants such that the denominator is zero, but the numerator is not zero. Then the system has *no* solution. For example, in the two equations

$$2x + 3y = 6$$
$$4x + 6y = 7 \tag{5-1}$$

we find

$$x = \frac{\begin{vmatrix} 6 & 3 \\ 7 & 6 \end{vmatrix}}{\begin{vmatrix} 2 & 3 \\ 4 & 6 \end{vmatrix}} = \frac{36 - 21}{12 - 12} = \frac{15}{0}.$$

This has no solution, since we could write

$$4x + 6y = 2(2x + 3y) = 2(6) \neq 7,$$

so the system of equations is contradictory.

On the other hand, the nonhomogeneous case can give a numerator which is zero, as well as a denominator which is zero. In this case there is an infinite number of solutions which will satisfy the simultaneous equations. For example, in the two equations

$$2x + 3y = 6$$
$$4x + 6y = 12 \tag{5-2}$$

we can solve for x as

$$x = \frac{\begin{vmatrix} 6 & 3 \\ 12 & 6 \end{vmatrix}}{\begin{vmatrix} 2 & 3 \\ 4 & 6 \end{vmatrix}} = \frac{36 - 36}{12 - 12} = \frac{0}{0}.$$

This division is again undefined, but we note that given any x we can solve for y in the original equations, and that this value of x and y will satisfy both equations. Thus, if $x = 0$, then $y = 2$; or else, if $y = 0$, then $x = 3$.

We can look at these two possibilities for the nonhomogeneous case by graphing the individual equations as in Figure 5-1. Each of the equations (5-1) represents a straight line in the x–y plane, as shown.

In the nonhomogeneous case

$$2x + 3y = 6$$
$$4x + 6y = 7 \tag{5-1}$$

the two straight lines are parallel but do not coincide. Thus there is no combination of x and y which lies on both lines, or which satisfies both equations.

In the second nonhomogeneous case where

$$2x + 3y = 6$$
$$4x + 6y = 12 \tag{5-2}$$

we note that both equations plot into the same straight line. Hence any combination of x and y which lies on this straight line will satisfy both equations, resulting in an infinite number of solutions.

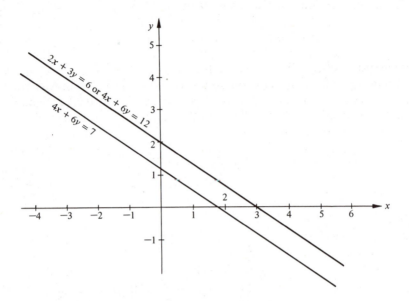

Figure 5-1.   Plotting of nonhomogeneous two-variable equations.

As to the homogeneous equations, we note that if we have the homogeneous equations

$$ax + by = 0$$
$$cx + dy = 0$$

and try to solve for x using the determinants, we have

$$x = \frac{\begin{vmatrix} 0 & b \\ 0 & d \end{vmatrix}}{\begin{vmatrix} a & b \\ c & d \end{vmatrix}} = \frac{0}{ad - bc}.$$

Thus the numerator is always zero in the homogeneous case. If the denominator $(ad - bc)$ is nonzero, then the above division holds and x, as well as y, is zero. In general, any system of n homogeneous equations whose determinant of coefficients is nonzero must have only the trivial solution that all the variables are zero.

But suppose that the denominator is zero as well. Then the division is not valid, and x and y need not be zero. This situation is then similar to the nonhomogeneous case where we have an infinite number of solutions. This is visible in Figure 5-2. If the denominator is nonzero, we have the case of

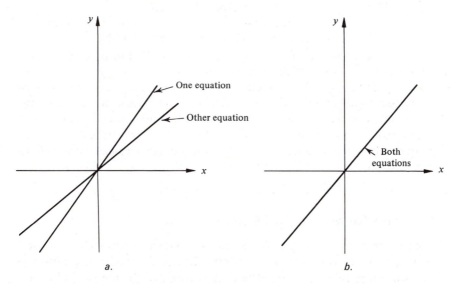

Figure 5-2. Plotting of homogeneous two-variable equations.

Figure 5-2a, where the two lines corresponding to the equation cross at the origin $x = y = 0$. If the denominator is zero as well, then the two straight lines coincide as in Figure 5-2b and there is an infinite number of solutions.

Let us now expand these ideas to the general case of n simultaneous equations, in n unknowns. To solve for the n unknowns, we express each as the quotient of two nth-order determinants. The denominator is simply the determinant of the coefficients, while the numerator is the same except that in solving for the value of x_i we replace the ith column by the set of y_i from the right of the equals signs.

As before, the set of equations can be either homogeneous or nonhomogeneous. In the nonhomogeneous case, there is no solution if the denominator (also called the *characteristic* determinant) is zero while the numerator is nonzero. If the numerator is zero, then there can be an infinite number of solutions. In the homogeneous case, on the other hand, the trivial solution $x_1 = x_2 = \cdots = x_n = 0$ will always work, and will in fact be the only one unless the characteristic determinant is zero, in which case there is an infinite number of solutions. In the third-order case we can easily explain these conclusions from examining the intersection of the planes defined by the equations, but this becomes progressively harder as the order rises.

Our results may be summarized this way:
　　1. Nonhomogeneous case
　　　　a. Denominator nonzero—a unique solution exists.
　　　　b. Denominator zero, numerator nonzero—no solution.
　　　　c. Both are zero—infinite number of solutions.

2. Homogeneous case (numerator always zero)
   a. Denominator nonzero—all variables are zero, trivial solution.
   b. Denominator zero—infinite number of solutions.

What about the difficulty of using Cramer's rule? For n equations in n unknowns, we must express each of the n variables x_1, x_2, ... x_n as the quotient of two nth-order determinants. In each case, the denominator is the same characteristic determinant, so that we have a total of n different numerator determinants, and one denominator determinant. Thus the complete solution of n equations requires us to evaluate $(n + 1)$ determinants of order n. In cases where n may be ten, twenty, or more, this becomes quite a chore. Hence Cramer's method is not too popular except where n is small.

5.8 The Inverse of a Matrix

In Example 5-1 we noted that when the product of two square matrices gives the identity (or unit) matrix, we say that the two matrices are inverses of each other. In general, we denote the inverse of the matrix $[A]$ by the symbol $[A]^{-1}$, so that we have

$$[A][A]^{-1} = [A]^{-1}[A] = [I]$$

(the fact that the multiplication can be done in either order is not too obvious, but true nevertheless).

Using determinants and the idea of cofactors (signed minors) we can express the inverse of a square matrix $[A]$ (if it exists) as

$$[A]^{-1} = \frac{1}{|A|} \begin{bmatrix} |A_{11}| & |A_{21}| & |A_{31}| & \cdots & |A_{n1}| \\ |A_{12}| & |A_{22}| & |A_{32}| & \cdots & |A_{n2}| \\ \vdots & & & & \vdots \\ |A_{1n}| & |A_{2n}| & |A_{3n}| & \cdots & |A_{nn}| \end{bmatrix}.$$

Alternatively, we can say that the inverse matrix has elements α_{ij} such that

$$\alpha_{ij} = \frac{|A_{ji}|}{|A|}.$$

The interchange of subscripts is important here.

Of course, it is quite clear that an inverse exists only if the determinant $|A|$ of the matrix is not zero, since otherwise each term in the inverse matrix would be infinite. In proper language, we say that a matrix whose determinant is equal to zero is *singular*, so that a square matrix has an inverse only if it is not singular. This is quite a logical conclusion, since the equation

$$[A][A]^{-1} = [I],$$

where $[I]$ is nonzero is impossible if $[A]$ is zero; otherwise it is somewhat similar to writing

$$0 \cdot \infty = 1,$$

which is mathematically taboo.

As an example, the matrix

$$[A] = \begin{bmatrix} 3 & 4 \\ 2 & 3 \end{bmatrix}$$

has a determinant whose value is 1, so that the inverse is

$$[A]^{-1} = \frac{1}{1} \begin{bmatrix} 3 & -4 \\ -2 & 3 \end{bmatrix} = \begin{bmatrix} 3 & -4 \\ -2 & 3 \end{bmatrix}.$$

Since each of the terms in the inverse is defined in terms of a ratio of two determinants, we could easily find the inverse for any matrix, if it exists. Nevertheless, for large matrices this may turn out to involve much work. For example, consider an $n \times n$ square matrix. To evaluate it, we need to evaluate the nth-order determinant of the matrix, as well as n^2 determinants of order $(n - 1)$, which represent the n^2 cofactors. The program of Example 5-2 could be used to find $|A|$, and with minor changes to omit certain rows and columns, could also be used to find the n^2 cofactors.

5.9 Use of Inverses to Solve Simultaneous Linear Equations

Assuming, for the moment, that finding the inverse of a large matrix is easy, we arrive at last at the crux of this chapter: solving simultaneous linear equations.

Let us again consider the case of n simultaneous equations in n unknowns $x_1, x_2, x_3, \ldots, x_n$, which are written as follows:

$$
\begin{aligned}
a_{11}x_1 + a_{12}x_2 + a_{13}x_3 + a_{14}x_4 + \cdots + a_{1n}x_n &= y_1 \\
a_{21}x_1 + a_{22}x_2 + a_{23}x_3 + a_{24}x_4 + \cdots + a_{2n}x_n &= y_2 \\
a_{31}x_1 + a_{32}x_2 + a_{33}x_3 + a_{34}x_4 + \cdots + a_{3n}x_n &= y_3 \\
&\vdots \\
a_{n1}x_1 + a_{n2}x_2 + a_{n3}x_3 + a_{n4}x_4 + \cdots + a_{nn}x_n &= y_n.
\end{aligned}
\tag{5-3}
$$

As before, we write the coefficients into the matrix

$$[A] = \begin{bmatrix} a_{11} & a_{12} & a_{13} & a_{14} & \cdots & a_{1n} \\ a_{21} & a_{22} & a_{23} & a_{24} & \cdots & a_{2n} \\ a_{31} & a_{32} & a_{33} & a_{34} & \cdots & a_{3n} \\ \vdots & & & & & \vdots \\ a_{n1} & a_{n2} & a_{n3} & a_{n4} & \cdots & a_{nn} \end{bmatrix}.$$

Let us now also define an $n \times 1$ column matrix called $[X]$ and another $n \times 1$ column matrix called $[Y]$ as follows

$$[X] = \begin{bmatrix} x_1 \\ x_2 \\ x_3 \\ \vdots \\ x_n \end{bmatrix} \qquad\qquad [Y] = \begin{bmatrix} y_1 \\ y_2 \\ y_3 \\ \vdots \\ y_n \end{bmatrix}.$$

Because of the very special way in which we defined matrix multiplication, we could write the original set of equations (5-3) exactly as

$$[A][X] = [Y].$$

Thus, for example, the first term in $[Y]$ (the term y_1) can be obtained by a term-by-term multiplication of the first row of $[A]$ (containing the terms $a_{11}, a_{12}, a_{13}, \ldots, a_{1n}$) by the first (and only) column of $[X]$ (containing the terms $x_1, x_2, x_3, \ldots, x_n$), and adding the partial products. This is exactly the same as writing the first equation

$$a_{11}x_1 + a_{12}x_2 + a_{13}x_3 + \cdots + a_{1n}x_n = y_1.$$

The form $[A][X] = [Y]$ is very similar to the form we would expect in a single-equation case such as $ax = y$. There we would simply find x by dividing

$$ax = y$$
$$x = y/a.$$

We cannot find $[X]$ in this case by solving

$$[X] = \frac{[Y]}{[A]},$$

since division is not defined for matrices. But we can do the next best thing—multiply by the inverse of $[A]$:

$$[A][X] = [Y]$$
$$[A]^{-1}[A][X] = [A]^{-1}[Y]$$
$$[X] = [A]^{-1}[Y].$$

Thus the problem of solving the set of n simultaneous equations is finished once we have found the inverse of the matrix of coefficients.

For example, consider the set of equations

$$3x_1 + 4x_2 = y_1$$
$$2x_1 + 3x_2 = y_2,$$

where y_1 might equal 11, and y_2 might equal 8. In matrix notation, we can write this set of equations as

$$\begin{bmatrix} 3 & 4 \\ 2 & 3 \end{bmatrix} \begin{bmatrix} x_1 \\ x_2 \end{bmatrix} = \begin{bmatrix} y_1 \\ y_2 \end{bmatrix}.$$

We found the inverse of the coefficient matrix earlier as

$$\begin{bmatrix} 3 & 4 \\ 2 & 3 \end{bmatrix}^{-1} = \begin{bmatrix} 3 & -4 \\ -2 & 3 \end{bmatrix},$$

so we could write

$$\begin{bmatrix} x_1 \\ x_2 \end{bmatrix} = \begin{bmatrix} 3 & -4 \\ -2 & 3 \end{bmatrix} \begin{bmatrix} y_1 \\ y_2 \end{bmatrix},$$

which can also be written as

$$x_1 = 3y_1 - 4y_2$$
$$x_2 = -2y_1 + 3y_2.$$

If we substitute the values of $y_1 = 11$ and $y_2 = 8$ from the original equations, we find that

$$x_1 = 33 - 32 = 1$$
$$x_2 = -22 + 24 = 2,$$

which, as we can easily prove, satisfy the original set of equations.

Just as in using Cramer's method, if the characteristic determinant, the determinant of the coefficients, is zero, we cannot use the method since in this case we cannot find the inverse.

Now what about the amount of work needed to produce a solution? With some careful reasoning, we see that the total effort needed is the same as for Cramer's method, since the evaluation of the inverse for an nth-order matrix requires us to evaluate n^2 determinants of order $(n - 1)$ and one determinant of order n, which is no easy feat.

But there is one particular case where the use of the inverse can mean a great saving. Suppose that we have a set of n simultaneous equations to be solved for different values of the $y_1, y_2, \ldots, y_n$. Using Cramer's method, each time we change the y_i we must go through the entire process to find the new values of the x_i. But if the coefficients $[A]$ stay the same, we need to find the inverse matrix only once, and then can use just a single matrix multiplication to find the x_i from the new y_i.

5.10 Gaussian Elimination for Simultaneous Linear Equations

By a coincidence, the method of Gaussian elimination which we have already examined in connection with evaluating determinants is also useful in solving linear simultaneous equations.

The simplest method for solving such equations, taught in high school, is in fact a Gaussian elimination of sorts. Consider, for example, the following set of three simultaneous equations:

$$\begin{aligned} 2x + y + z &= 7 \\ 4x + 4y + 3z &= 21 \\ 6x + 7y + 4z &= 32. \end{aligned} \qquad (5\text{-}4)$$

Suppose we multiply the first equation by 2 and subtract from the second equation, and also multiply the first equation by 3 and subtract from the third equation. Then we obtain

$$\begin{aligned} 2x + y + z &= 7 \\ 2y + z &= 7 \\ 4y + z &= 11. \end{aligned}$$

Next we multiply the second equation by 2 and subtract the result from the third equation to obtain

$$\begin{aligned} 2x + y + z &= 7 \\ 2y + z &= 7 \\ -z &= -3. \end{aligned}$$

This is exactly the same as triangularizing a determinant. Once we have gone this far, we now proceed through the equations, one by one, starting from the bottom, to find the values of x, y, and z; this process is called *back substitution*.

From the bottom equation, we see that $z = 3$. We now substitute this value into the second equation and solve for $y = 2$. Finally, we substitute z and y into the first equation and solve for $x = 1$, and we are finished.

This exact procedure could have been done as an array of numbers. Start with the array of coefficients

$$\begin{bmatrix} 2 & 1 & 1 \\ 4 & 4 & 3 \\ 6 & 7 & 4 \end{bmatrix} = [A]$$

and add a fourth column which consists of the values from the right of the equals signs in Eqs. (5-4):

$$\begin{bmatrix} 7 \\ 21 \\ 32 \end{bmatrix} = [Y].$$

Put these alongside each other to obtain the so-called *augmented matrix*

$$\begin{bmatrix} 2 & 1 & 1 & 7 \\ 4 & 4 & 3 & 21 \\ 6 & 7 & 4 & 32 \end{bmatrix}.$$

From now on we perform a triangularization, as in Gaussian elimination, except for two small changes. First, we have a rectangular matrix instead of a square determinant, and so there is really no main diagonal. Hence the result will not be really triangular since there are more columns than rows. Further, if we switch two rows this is equivalent to switching two of the original equations, and in no way changes the answer. Thus we need not keep track of row interchanges for the purpose of deciding on the sign to be used.

In triangularizing the matrix we are going to interchange rows, and subtract multiples of one row from another. We never defined these operations as being valid for matrices, and they are not. But notice that, *in this particular case*, we are entitled to use them since they are valid for the original equations, and the matrix notation we are using is only a means of shorthand notation for the original equations.

As before, we start with the augmented matrix

$$\begin{bmatrix} 2 & 1 & 1 & 7 \\ 4 & 4 & 3 & 21 \\ 6 & 7 & 4 & 32 \end{bmatrix} \qquad (5\text{-}5)$$

and multiply the first row by 2 and 3, respectively, to subtract from the second and third rows:

$$\begin{bmatrix} 2 & 1 & 1 & 7 \\ 0 & 2 & 1 & 7 \\ 0 & 4 & 1 & 11 \end{bmatrix}.$$

Then we multiply the second row by 2 and subtract from the third:

$$\begin{bmatrix} 2 & 1 & 1 & 7 \\ 0 & 2 & 1 & 7 \\ 0 & 0 & -1 & -3 \end{bmatrix}.$$

Finally, we perform the back-substitution to find the three unknowns. Let us now extend the method to n equations in n unknowns. Suppose that the n equations are written, as before,

$$a_{11}x_1 + a_{12}x_2 + a_{13}x_3 + \cdots + a_{1n}x_n = y_1$$
$$a_{21}x_1 + a_{22}x_2 + a_{23}x_3 + \cdots + a_{2n}x_n = y_2$$
$$a_{31}x_1 + a_{32}x_2 + a_{33}x_3 + \cdots + a_{3n}x_n = y_3$$
$$\vdots$$
$$a_{n1}x_1 + a_{n2}x_2 + a_{n3}x_3 + \cdots + a_{nn}x_n = y_n.$$

To evaluate, we form the augmented matrix

$$\begin{bmatrix} a_{11} & a_{12} & a_{13} & \cdots & a_{1n} & y_1 \\ a_{21} & a_{22} & a_{23} & \cdots & a_{2n} & y_2 \\ a_{31} & a_{32} & a_{33} & \cdots & a_{3n} & y_3 \\ \vdots & & & & \vdots & \\ a_{n1} & a_{n2} & a_{n3} & \cdots & a_{nn} & y_n \end{bmatrix}.$$

After triangularization, we have the new matrix (unless some rows have been interchanged)

$$\begin{bmatrix} a_{11} & a_{12} & a_{13} & \cdots & a_{1n} & y_1 \\ 0 & a'_{22} & a'_{23} & \cdots & a'_{2n} & y'_2 \\ 0 & 0 & a''_{33} & \cdots & a''_{3n} & y''_3 \\ \vdots & & & & & \vdots \\ 0 & 0 & 0 & \cdots & a_{nn}^{(n-1)} & y_n^{(n-1)} \end{bmatrix}.$$

Finally, we perform the back-substitution, and we have the answer. The following example will make this process clearer.

5.11 Example 5-3. Solution of Simultaneous Equations by Gauss Elimination

In this example, we shall solve the following set of four simultaneous equations:

$$-x_1 + 3x_2 + 5x_3 + 2x_4 = 10$$
$$x_1 + 9x_2 + 8x_3 + 4x_4 = 15$$
$$x_2 \quad\quad + x_4 = 2$$
$$2x_1 + x_2 + x_3 - x_4 = -3.$$

The characteristic determinant of this system is

$$\begin{vmatrix} -1 & 3 & 5 & 2 \\ 1 & 9 & 8 & 4 \\ 0 & 1 & 0 & 1 \\ 2 & 1 & 1 & -1 \end{vmatrix} = 14,$$

so that a solution exists for this nonhomogeneous set, and in fact, the solution is $x_1 = -1$, $x_2 = 0$, $x_3 = 1$, and $x_4 = 2$.

The resulting augmented matrix is

$$\begin{bmatrix} -1 & 3 & 5 & 2 & 10 \\ 1 & 9 & 8 & 4 & 15 \\ 0 & 1 & 0 & 1 & 2 \\ 2 & 1 & 1 & -1 & -3 \end{bmatrix}.$$

The first part of the following program is the same as the elimination portion of the program of Example 5-2, except that it is adapted to handle a 4×5 matrix. As before, as each succeeding column is reduced, we look for the largest pivot. If no nonzero pivot is found, the program exits at statement 60. In Example 5-2, the absence of a nonzero pivot indicated that the determinant had a value of zero; here too, the absence of a nonzero pivot indicates that the characteristic determinant is zero, so that no unique solution exists.

Once the reduced system of equations is found, the back substitution is done by letting

$$x_4 = y_4/a_{44},$$

since all other coefficients in the fourth row are zero; then

$$x_3 = (y_3 - a_{34}x_4)/a_{33},$$

since the other coefficients in the third row are zero; and similarly,

$$x_2 = (y_2 - a_{24}x_4 - a_{23}x_3)/a_{22}$$
$$x_1 = (y_1 - a_{14}x_4 - a_{13}x_3 - a_{12}x_2)/a_{11}.$$

Rather than write out each of these equations explicitly, we write a loop consisting of statements 350 through 500 which performs the calculations. Since in a DO loop the index must increase, we define a new index IREV = M + 1 − I which starts at M and goes back to 1. Thus we start on the last row, and work our way back up. Each row except the last requires several subtractions, which are done in statements 400 through 450, and the final division by a_{ii} is done in statement 500. Finally, after all unknowns are found, they are printed and the program is over.

The following program is annotated and is worth some study.

```
      DIMENSION A(4, 5), X(5)
      M = 4
      N = 5
C     READ TERMS FROM CARD
      READ (2, 10) ((A(I, J), J = 1, 5), I = 1, 4)
   10 FORMAT (4(5F5.0, /))
      LAST = M - 1
C     START OVERALL LOOP FOR (M-1) PIVOTS
      DO 200 I = 1, LAST
C
C     FIND THE LARGEST REMAINING TERM IN I-TH COLUMN FOR PIVOT
C
      BIG = 0.
      DO 50 K = I, M
      TERM = ABS (A(K, I))
      IF (TERM - BIG) 50, 50, 30
   30 BIG = TERM
      L = K
   50 CONTINUE
```

```
C       CHECK WHETHER A NON-ZERO TERM HAS BEEN FOUND
        IF (BIG) 80, 60, 80
   60 CALL EXIT
C       L-TH ROW HAS THE BIGGEST TERM -- IS I = L
   80 IF (I - L) 90, 120, 90
C       I IS NOT EQUAL TO L, SWITCH ROWS I AND L
   90 DO 100 J = 1, N
        TEMP = A(I, J)
        A(I, J) = A(L, J)
  100 A(L, J) = TEMP
C
C       NOW START PIVOTAL REDUCTION
C
  120 PIVOT = A(I, I)
        NEXTR = I + 1
C       FOR EACH OF THE ROWS AFTER THE I-TH
        DO 200 J = NEXTR, M
C       MULTIPLYING CONSTANT FOR THE J-TH ROW IS
        CONST = A(J, I) / PIVOT
C       NOW REDUCE EACH TERM OF THE J-TH ROW
        DO 200 K = I, N
  200 A(J, K) = A(J, K) - CONST * A(I, K)
C       END OF PIVOTAL REDUCTION - PRINT NEW MATRIX
        WRITE (3, 320) ((A(I, J), J = 1, N), I = 1, M)
  320 FORMAT (5F12.6, /)
C
C       PERFORM BACK SUBSTITUTION
C
  350 DO 500 I = 1, M
C       IREV IS THE BACKWARD INDEX, GOING FROM M BACK TO 1
        IREV = M + 1 - I
C       GET Y(IREV) IN PREPARATION
        Y = A(IREV, N)
        IF (IREV - M) 400, 500, 400
C       NOT WORKING ON LAST ROW, I IS 2 OR GREATER
  400 DO 450 J = 2, I
C       WORK BACKWARD FOR X(N), X(N-1) ----, SUBSTITUTING
C       PREVIOUSLY FOUND VALUES
        K = N + 1 - J
  450 Y = Y - A(IREV, K) * X(K)
C       FINALLY, COMPUTE X(IREV)
  500 X(IREV) = Y / A(IREV, IREV)
C
C       PRINT VALUES OF X(K)
C
        DO 600 I = 1, M
        WRITE (3, 550) I, X(I)
  550 FORMAT (' X(', I2, ') = ', F10.6)
  600 CONTINUE
        CALL  EXIT
        END
```

The results are:

```
    2.000000    1.000000    1.000000   -1.000000   -3.000000

    0.000000    8.500001    7.500000    4.500000   16.500003

    0.000000    0.000000    2.411765   -0.352941    1.705882

    0.000000    0.000000   -0.000000    0.341463    0.682927

 X( 1) =   -1.000000
 X( 2) =    0.000000
 X( 3) =    0.999999
 X( 4) =    1.999999
```

Thus the computer gives us the reduced (triangularized) system of equations as

$$2.000000x_1 + 1.000000x_2 + 1.000000x_3 - 1.000000x_4 = -3.000000$$
$$8.500001x_2 + 7.500000x_3 + 4.500000x_4 = 16.500003$$
$$2.411765x_3 - 0.352941x_4 = 1.705882$$
$$0.341463x_4 = 0.682927$$

from which it obtains the (almost) correct set of answers. The errors are caused by roundoff errors, and are quite small. Since there are only 4 equations to be solved, we find that double precision arithmetic would have given us the exact answers of $-1, 0, 1$, and 2, respectively.

5.12 Ill-conditioned Systems and Other Errors

As already pointed out, roundoff error plays a large part in the solution of sets of simultaneous linear equations, mainly because of the very large number of calculations to be performed. There is an especially great effect on pivotal condensation (Gaussian elimination) because each result depends on all of the previous results. Thus an error in one of the early steps tends to cause other errors later—hence we have propagation errors.

Since we know that we shall have errors in the answer, the question remains whether we have any way of finding out just how large these errors might be. One possible way of checking this is to substitute the solutions we obtain back into the simultaneous equations, and thereby see whether they are really satisfied.

Consider, for example, the two simultaneous equations

$$3x_1 + 4x_2 = 7$$
$$5x_1 - 2x_2 = 3.$$

Here quite obviously $x_1 = x_2 = 1$. Suppose, however, that a slight error in our calculations results with

$$x_1 = 0.999$$
$$x_2 = 1.002.$$

Substituting these calculated values back into the two equations gives us

$$3(0.999) + 4(1.002) = 7.005$$
$$5(0.999) - 2(1.002) = 2.991.$$

Comparing these with the values of 7 and 3, respectively, we are tempted to think that, since there is an error in the third decimal place, x_1 and x_2 are both accurate to about two decimal places. And, true enough, many times we may be right.

But look at the following two equations:

$$x_1 + x_2 = 2$$
$$1.01x_1 + x_2 = 2.01. \tag{5-6}$$

Here again we see that $x_1 = x_2 = 1$ is an exact answer. Yet suppose that, due to error, we calculated

$$x_1 = 0$$
$$x_2 = 2.005.$$

Substituting back into the two equations, we have

$$(0) + (2.005) = 2.005$$
$$1.01(0) + (2.005) = 2.005.$$

Comparing the values of 2 and 2.01, respectively, we see an error of only 0.005, yet the error in each answer is about 1.

But there is actually a second possible source of error. Suppose that the coefficients themselves are in error, as could easily happen. Thus, suppose that some error has changed equations (5-6) into

$$x_1 + x_2 = 2$$
$$1.0001x_1 + x_2 = 2.007.$$

Even an utterly ridiculous solution such as $x_1 = 100$, $x_2 = -98$ will fit the changed equations quite closely. Such sets of equations that are *approximately* satisfied even by completely wrong answers are called *ill-conditioned*.

The reason for the ill-conditioning is the fact that the determinant of the coefficients is almost zero. Since we have already found that a zero characteristic determinant leads (in the case of nonhomogeneous equations) to either an infinite number of solutions or to no solution at all, a near-zero determinant is likely to be bad also.

Suppose we have two equations in two unknowns,

$$a_{11}x_1 + a_{12}x_2 = y_1$$
$$a_{21}x_1 + a_{22}x_2 = y_2,$$

and solve for x_1 in terms of x_2 in each equation:

$$x_1 = (y_1 - a_{12}x_2)/a_{11} \qquad \text{from the first equation}$$

$$x_1 = \frac{y_1}{a_{11}} - \frac{a_{12}}{a_{11}}x_2 \tag{5-7}$$

$$x_1 = (y_2 - a_{22}x_2)/a_{21} \qquad \text{from the second equation}$$

$$x_1 = \frac{y_2}{a_{21}} - \frac{a_{22}}{a_{21}}x_2. \tag{5-8}$$

Let us draw the straight lines represented by the two equations in the x_2–x_1 plane as in Figure 5-3.

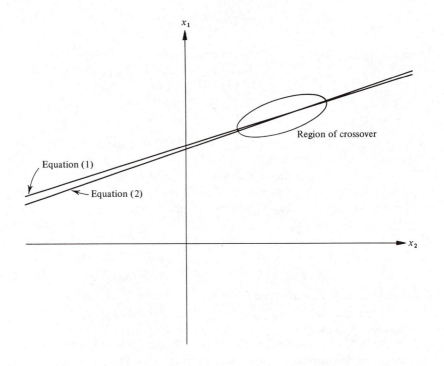

Figure 5-3.   An ill-conditioned second-order system.

The solution for the system is the value of x_1 and x_2 at the point of intersection of the two straight lines. If, however, the two lines are almost parallel, then it is hard to tell just where the lines really cross.

This occurs when the slopes of the two lines are nearly equal. Having expressed the equations (5-7) and (5-8) for the two lines in terms of $x_1 = f(x_2)$, we need only look at the coefficient of the x_2 term—we have really the case of $y = mx + b$, where the number m is the slope, and b is the y-intercept. Thus the ill-conditioned case occurs when the two slopes are nearly equal, that is,

$$a_{12}/a_{11} \approx a_{22}/a_{21}$$

or

$$a_{22}a_{11} \approx a_{12}a_{21}$$

or, indeed, if

$$a_{22}a_{11} - a_{12}a_{21} \approx 0.$$

But this is the case when $|A|$, the characteristic determinant, is near zero, since $|A| = a_{22}a_{11} - a_{12}a_{21}$.

Thus, an ill-conditioned system is indicated when the characteristic determinant is near zero, since even a very slight error in specifying the two equations is likely to move the actual intersection to a completely different place. While the preceding explanation is in terms of a second-order system, obviously we can generalize to more equations in more unknowns, to say that the value of the determinant indicates whether a system is ill-conditioned or not.

But, just how "near zero" may the determinant be before we have trouble? Since a system of simultaneous equations can be multiplied through by a scale factor, we can change each of the terms in the characteristic determinant without necessarily changing the solutions. Thus the value of the determinant by itself need not indicate whether a system is ill-conditioned, since "near zero" is only relative.

One way of deciding whether the determinant is too small is as follows. For each row of the determinant, we determine a scale factor equal to the square root of the sum of the squares of the terms of that row. Thus, for example, for the ith row, the scale factor k_i would be

$$k_i = (a_{i1}^2 + a_{i2}^2 + a_{i3}^2 + \cdots + a_{in}^2)^{\frac{1}{2}}.$$

We then divide each element of that row by the scale factor. Repeating for each row, we obtain a normalized determinant, which we evaluate. If the value of this normalized determinant is substantially smaller than 1 in magnitude, we say that the system of equations is ill-conditioned.

While perfectly valid, the above method requires the solution of the normalized characteristic determinant, a process which would normally not be done. For a large system of equations, this may take time.

Another method, somewhat simpler though even more lengthy, is based on the ideas shown by Figure 5-3. In the case of two equations in two unknowns, an ill-conditioned system is indicated when the two lines representing the two equations are nearly parallel. If one of the lines is displaced slightly, as a result perhaps of a slight change of the coefficients of the system, the point of intersection will change by a large amount; the same conclusion will hold for systems of more than two equations. To test for ill-conditioning, we may therefore change the coefficients slightly, repeat the entire solution, and compare the two solutions obtained.

If the two solutions are different by a much larger magnitude than the change made in the coefficients, we may suspect ill-conditioning. In that case, we should probably pause, and reconsider whether we wish to trust the computer solution at all.

5.13 Gauss-Seidel Iterative Method

In large systems of simultaneous equations, and especially in ill-conditioned systems, roundoff error can destroy the accuracy of the solution, since it is propagated all along the solution. In such cases we may decide to use an iterative method, which has the advantage of not being too sensitive to roundoff error, provided that it converges.

In Chapter 4 we discussed iterative methods for the solution of nonlinear simultaneous equations; these methods are perfectly suitable for linear equations as well. They have the advantage that roundoff error does not propagate; on the other hand, they have the disadvantage that they may not converge at all, and that they generally require many more computations than the direct matrix methods discussed earlier in this chapter. The Gauss-Seidel method is such an iterative method, designed specifically for linear equations; it is especially useful if the characteristic determinant has many zero terms.

As before, let us consider the n equations

$$a_{11}x_1 + a_{12}x_2 + a_{13}x_3 + \cdots + a_{1n}x_n = y_1$$
$$a_{21}x_1 + a_{22}x_2 + a_{23}x_3 + \cdots + a_{2n}x_n = y_2$$
$$a_{31}x_1 + a_{32}x_2 + a_{33}x_3 + \cdots + a_{3n}x_n = y_3$$
$$\vdots$$
$$a_{n1}x_1 + a_{n2}x_2 + a_{n3}x_3 + \cdots + a_{nn}x_n = y_n.$$

where we assume that the diagonal terms $a_{11}, a_{22}, a_{33}, \ldots, a_{nn}$ are all nonzero; if necessary, the equations must be rearranged so that all of these terms are nonzero.

We now solve the first equation for x_1, solve the second for x_2, and so on, to get

$$x_1 = \frac{1}{a_{11}}(y_1 - a_{12}x_2 - a_{13}x_3 - \cdots - a_{1n}x_n)$$

$$x_2 = \frac{1}{a_{22}}(y_2 - a_{21}x_1 - a_{23}x_3 - \cdots - a_{2n}x_n)$$

$$x_3 = \frac{1}{a_{33}}(y_3 - a_{31}x_1 - a_{32}x_2 - \cdots - a_{3n}x_n)$$

$$\vdots$$

$$x_n = \frac{1}{a_{nn}}(y_n - a_{n1}x_1 - a_{n2}x_2 - \cdots - a_{n,n-1}x_{n-1}).$$

We now start the iteration by picking an initial set of guesses for all the x_i, inserting these into the first equation and calculating a new x_i. Then we insert this value of x_i, as well as the previously guessed values for $x_2, x_3, \ldots, x_n$ into the second equation and solve for x_2. We proceed down the list until we calculate a new value for x_n, and then return to the top and repeat.

This process, if it converges at all, is likely to take many iterations before reasonable accuracy results; for this reason it is generally used only if direct methods, such as the Gaussian elimination or solution by inverse, fail. One possible exception to this may occur if many of the coefficients of the set of equations are zero, as occasionally happens in certain problems. In this case the iterations may be relatively simple and fast since few multiplications and additions are needed.

The question of convergence is not simple, since it is often difficult to even guess how far to iterate before stopping. The criteria for stopping the iteration are usually

1. Either the number of iterations has exceeded some predetermined maximum k, or
2. The differences between successive values of all x_i are less than some predetermined value of ε. This particular criterion is tricky, because it may be satisfied if convergence is very, very slow, even though the answers are not anywhere near correct.

If the iteration appears to diverge, then (as we saw in Chapter 4) rearrangement of the equations may produce convergence after all. In this case two ways of rearranging the equations are possible:

1. Instead of solving the first equation for x_1, the second for x_2, and so on, we may decide to solve different equations for different x_i. The one restraint here is that the coefficient of the x_i being solved for must not be zero.

2. Even after a particular order is used for solving for different x_i, we may perform the iteration itself in a different order. Thus, instead of picking initial guesses, and then solving in order for $x_1, x_2, x_3, \ldots, x_n$, we might first solve for x_5, then for x_2, then for x_{13}, etc. For n equations, there are $n!$ different orders in which they may be solved, and one of these might work.

5.14 Example 5-4. Solution by Gauss-Seidel Iteration

As the last example in this chapter, we perform an iterative Gauss-Seidel solution of the four equations from Example 5-3:

$$-x_1 + 3x_2 + 5x_3 + 2x_4 = 10$$
$$x_1 + 9x_2 + 8x_3 + 4x_4 = 15$$
$$x_2 \qquad + \ x_4 = 2$$
$$2x_1 + \ x_2 + \ x_3 - \ x_4 = -3.$$

Since we cannot solve the third equation for x_3, we solve for x_1 in the first equation, for x_2 in the second, for x_4 in the third, and for x_3 in the fourth, so that we have

$$x_1 = (10 - 3x_2 - 5x_3 - 2x_4)/(-1)$$
$$x_2 = (15 - x_1 - 8x_3 - 4x_4)/9$$
$$x_3 = (-3 - 2x_1 - x_2 + x_4)/1$$
$$x_4 = (2 - x_2)/1;$$

this is the most obvious way of solving the equations, though not a feasible one. As we run the following program, we see that it diverges and, even after 100 iterations, produces results which are outside the limits imposed by the FORMAT statement:

```
      C       INITIALIZE GUESSES
              X1 = 0.
              X2 = 0.
              X3 = 0.
              X4 = 0.
      C       START ITERATION
              DO 100 I = 1, 100
              X1 = (10 - 3. * X2 - 5. * X3 - 2. * X4) / (-1)
              X2 = (15 - X1 - 8. * X3 - 4. * X4) / 9.
              X3 = (-3. - 2. * X1 - X2 + X4) / 1.
              X4 = (2. - X2) / 1.
      100 CONTINUE
      C       PRINT RESULTS
              WRITE (3, 200) X1, X2, X3, X4
      200 FORMAT (4F10.6)
              CALL EXIT
              END
```

The result is a row of asterisks, which shows that the answers must be out of range.

****

Now, suppose that we had solved for the x_i in a different order, like this:

Solve for x_1 in equation 4

,,   ,, x_2 ,,   ,,   2

,,   ,, x_3 ,,   ,,   1

,,   ,, x_4 ,,   ,,   3

We change the statements inside the iteration loop to read

```
X1 = (-3. - X2 - X3 + X4) / 2.
X2 = (15 - X1 - 8. * X3 - 4. * X4) / 9.
X3 = (10 + X1 - 3. * X2 - 2. * X4) / 5.
X4 = (2. - X2) / 1.
```

This time our answer is the set of numbers

```
-1.000001   0.000001   0.999999   1.999998
```

which, except for a slight roundoff error, are correct. Although 100 iterations were performed in this program, actually only 73 would have been needed for these answers. Still, this process takes longer than a direct method.

As described earlier, whether the iteration converges depends on how the equations are ordered, and which equation is solved for which x_i. To simplify this procedure, we mention here a simple theorem which provides sufficient, though not necessary, conditions for convergence:

Theorem *The Gauss-Seidel iteration will converge if, in the characteristic determinant, each term on the main diagonal is larger (in absolute value) than the sum of the absolute values of all the other terms in the same row or column. That is, we are guaranteed convergence if,*

$$|a_{ii}| > \sum_{\substack{j=1 \\ j \neq i}}^{n} |a_{ij}| \qquad \text{for all } i = 1, 2, \ldots, n$$

and

$$|a_{ii}| > \sum_{\substack{j=1 \\ j \neq i}}^{n} |a_{ji}| \qquad \text{for all } i = 1, 2, \ldots, n.$$

The theorem is not very useful unless we can somehow rearrange the equations to satisfy the theorem as closely as we can; this is not altogether easy. When we rearranged the equations above, we reordered them to produce this system:

$$\text{Solve for } x_1: \quad 2x_1 + x_2 + x_3 - x_4 = -3$$
$$\text{Solve for } x_2: \quad x_1 + 9x_2 + 8x_3 + 4x_4 = 15$$
$$\text{Solve for } x_3: \quad -x_1 + 3x_2 + 5x_3 + 2x_4 = 10$$
$$\text{Solve for } x_4: \quad x_2 \qquad + x_4 = 2.$$

In this case, we arranged the equations to make the terms on the main diagonal as large as we can. It usually happens, as here, that we cannot fulfil the conditions of the theorem fully, but since the theorem is not necessary for convergence even some improvement may help.

5.15 Problems

1. For the following two matrices $[A]$ and $[B]$, perform the following operations:

a. Find their sum $[A] + [B]$

b. Find their difference $[A] - [B]$

c. Find their product $[A][B]$

d. Find their product $[B][A]$ and compare with $[A][B]$

e. Find their inverses $[A]^{-1}$ and $[B]^{-1}$

f. Calculate

$$[A][I]$$
$$[A][A]^{-1}$$
$$[A]^{-1}[A]$$
$$[B][B]^{-1}$$
$$\{[A][B]\}[B]^{-1}$$
$$[B]^{-1}[B]$$

$$[A] = \begin{bmatrix} 1 & 2 & 3 \\ 3 & 3 & 5 \\ 0 & 2 & 3 \end{bmatrix} \quad [B] = \begin{bmatrix} 7 & 6 & 5 \\ 5 & 4 & 3 \\ 3 & 7 & 6 \end{bmatrix}.$$

· **2.** Manually evaluate the following determinant by two different methods:

$$\begin{vmatrix} 1 & 2 & 3 & 4 \\ 0 & 1 & 2 & 3 \\ 1 & 2 & 4 & 6 \\ 1 & 2 & 4 & 7 \end{vmatrix}.$$

3. Solve the following equations for x, y, and z using Cramer's rule and also Gaussian elimination,

$$x + 3y - z = 4$$
$$5x - 2y - z = -2$$
$$2x + 2y + z = 9.$$

4. Using the program of Example 5-2 to evaluate a determinant, write a new program which will solve the simultaneous equations of problem 3 by Cramer's rule.

5. In Examples 5-2 and 5-3, we triangularize a matrix by pivotal condensation, with the result that the terms below the main diagonal become zero. Knowing in advance that these terms should be zero, there is really no need to actually compute them; in fact, doing so is just a rather large waste of time. Modify the programs of Examples 5-2 and 5-3 to eliminate this extra computation.

6. Using two methods discussed in this chapter, determine whether the following equations represent an ill-conditioned system:

$$20x + 99y = 218$$
$$101x + 500y = 1101.$$

7. Plot the two lines represented by the equations in problem 6, and explain your answer to problem 6.

8. Write a program to find the inverse of a 5×5 matrix.

9. Rearrange the equations of problem 3 in such a way that the Gauss-Seidel method is most likely to converge; then try a hand iteration to see whether your method works, starting with initial values of $x = y = z = 0$. When convergence seems to be occurring, use the program of Example 5-4 to find the roots.

10. Without too much additional effort, the program of Example 5-3 could have been modified to calculate the value of $|A|$. Make this change.

6

Numerical Integration

There are essentially two different types of integrals: indefinite integrals and definite integrals.

A typical indefinite integral is

$$\int x^2 \, dx = \frac{x^3}{3}.$$

The answer to this integration is a function, in this case a function of x.

A typical definite integral is

$$\int_0^1 x^2 \, dx = \tfrac{1}{3}.$$

The answer to this integration is a number, in this case $\tfrac{1}{3}$.

Since the solution of an indefinite integral is a function and not a mere number, it must be obtained by calculus and not by numerical methods. Although some research has been done to permit computer solution of indefinite integrals, this problem has not been sufficiently solved to date.

The evaluation of definite integrals, on the other hand, can be done by computer, and we shall show several methods for doing so in this chapter.

6.1 Introduction to Numerical Integration

From your knowledge of calculus, you know that the evaluation of a definite integral such as

$$\int_0^1 x^2 \, dx$$

is done in two steps. First, we interpret this integral as an indefinite integral, and integrate to get

$$\int x^2 \, dx = \frac{x^3}{3}.$$

In order to now find the definite integral between the limits of 0 and 1, we evaluate the function $x^3/3$ between the limits:

$$\int_0^1 x^2 \, dx = \left[\frac{x^3}{3}\right]_0^1 = \left(\frac{1^3}{3}\right) - \left(\frac{0^3}{3}\right) = \frac{1}{3}.$$

But since this requires us to evaluate an indefinite integral first, this method is obviously not suitable for computer solution.

Instead, we note that the above integral can be interpreted as the area under the curve $y = x^2$ between the limits of $x = 0$ and $x = 1$, as shown in Figure 6-1.

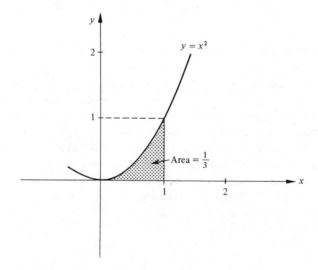

Figure 6-1. $\int_0^1 x^2 \, dx$ is equal to the shaded area under the curve $y = x^2$.

To evaluate a definite integral we therefore find the area under the corresponding curve. This process is done by a numerical approximation procedure, and is called *numerical integration*; it is also often called *quadrature*.

6.2 The Trapezoidal Method

The simplest (though far from most accurate) method of finding the area under a curve is by approximating that area with a series of trapezoids.

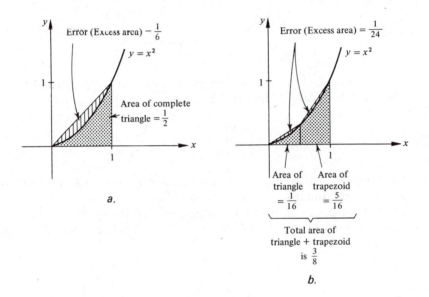

Figure 6-2. Approximation of an area by a number of trapezoids.

Figure 6-2*a* shows how we might approximate the area under the curve $y = x^2$ with a triangle (a triangle is a degenerate case of a trapezoid). Obviously we make quite an error because the triangle contains a far larger area than the actual integral, and so we consider breaking up the area into two trapezoids, as in Figure 6-2*b*. The error is now smaller, and so we consider using three, four, or more trapezoids for even better accuracy. Theoretically, using an infinite number of trapezoids should give perfect accuracy, but roundoff error will give us problems as we shall soon see.

In general, suppose we wish to integrate the function $f(x)$ shown in Figure 6-3 between the limits of a and b:

$$I = \int_a^b f(x)\, dx.$$

To use the trapezoidal rule, we divide the interval from a to b into n equal parts as shown. The boundaries of the trapezoids are the values x_0, x_1, x_2, ..., x_n.

Now consider the ith trapezoid, which lies between x_{i-1} and x_i. Its width $w = (b - a)/n$, the height of its left side is $f(x_{i-1})$, and the height of its right side is $f(x_i)$. Its area is therefore

$$A_i = \frac{w}{2}[f(x_{i-1}) + f(x_i)].$$

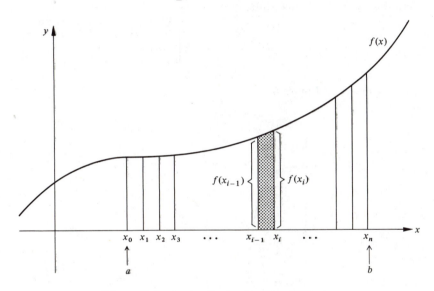

Figure 6-3. Integration by the trapezoidal rule.

The total area of all n trapezoids is the trapezoidal approximation to the integral, and is called T_n:

$$T_n = A_1 + A_2 + A_3 + \cdots + A_{n-1} + A_n$$

$$= \frac{w}{2}[f(x_0) + f(x_1)] + \frac{w}{2}[f(x_1) + f(x_2)] + \frac{w}{2}[f(x_2) + f(x_3)]$$

$$+ \cdots + \frac{w}{2}[f(x_{n-2}) + f(x_{n-1})] + \frac{w}{2}[f(x_{n-1}) + f(x_n)]$$

$$= \frac{w}{2}[f(x_0) + 2f(x_1) + 2f(x_2) + \cdots + 2f(x_{n-1}) + f(x_n)] \qquad (6\text{-}1)$$

As pointed out earlier, the accuracy in the trapezoidal method depends on the number n of trapezoids we take. Obviously the area under a straight line can be obtained with only one trapezoid, while the area under a rapidly changing curve with many bends may require many trapezoids. Theoretically we wish many trapezoids, but for practical reasons of speed and low roundoff error a small number may be better, as we shall see in the following example.

6.3 Example 6-1. Integration with the Trapezoidal Method

In this example we shall integrate

$$\int_0^1 (6 - 6x^5)\, dx = \left[6x - x^6 \right]_0^1 = (6 - 1) = 5,$$

which is shown in Figure 6-4.

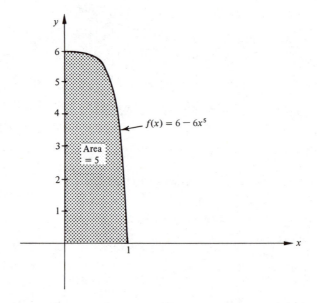

Figure 6-4.   The integral of Example 6-1.

For illustrative purposes, our program will use 1, 2, 4, 8, 16, ..., 16348 trapezoids, so that we may compare results for different numbers of trapezoids. For each trapezoid, the program calculates XL and XR, the left and right boundaries respectively, and the area A of that trapezoid. This area is then added into the total AREA at the end of the inner DO loop, and the total is printed after all the trapezoids are counted.

We use the following program

```
F(X) = 6. - 6. * X ** 5
      WRITE (3, 10)
10    FORMAT (' TRAPEZOIDS', 6X, 'AREA', 12X, 'ERROR')
      DO 50 I = 1, 15
      NODIV = 2 ** (I - 1)
      AREA = 0.
      WIDTH = 1. / NODIV
      DO 40 J = 1, NODIV
      XL = (J - 1) * WIDTH
      XR = J * WIDTH
      A = (WIDTH / 2.) * (F(XL) + F(XR))
40    AREA = AREA + A
      ERROR = 5. - AREA
50    WRITE (3, 60) NODIV, AREA, ERROR
60    FORMAT (I8, F16.8, E19.8)
      CALL EXIT
      END
```

and get the following results:

TRAPEZOIDS	AREA	ERROR
1	3.00000048	0.20000004E 01
2	4.40625096	0.59375012E 00
4	4.84570408	0.15429690E 00
8	4.96106053	0.38940437E-01
16	4.99023820	0.97627658E-02
32	4.99755193	0.24490361E-02
64	4.99937249	0.62847149E-03
128	4.99981881	0.18215182E-03
256	4.99990082	0.10013581E-03
512	4.99985410	0.14686587E-03
1024	4.99971677	0.28419500E-03
2048	4.99947263	0.52833568E-03
4096	4.99893094	0.10700228E-02
8192	4.99783326	0.21677021E-02
16384	4.99563695	0.43640146E-02

Looking over these results we note that the error seems to be decreasing until we reach 256 trapezoids, and then again increases. To see the effect of roundoff error on these results, let us repeat the program using double-precision arithmetic[1]:

TRAPEZOIDS	AREA	ERROR
1	3.00000000	0.20000000E 01
2	4.40625000	0.59375000E 00
4	4.84570312	0.15429687E 00
8	4.96105957	0.38940429E-01
16	4.99024200	0.97579956E-02
32	4.99755904	0.24409629E-02
64	4.99938962	0.61037764E-03
128	4.99984729	0.15271455E-03
256	4.99996161	0.38392841E-04
512	4.99998995	0.10050833E-04
1024	4.99999659	0.34086406E-05
2048	4.99999732	0.26822090E-05
4096	4.99999565	0.43511390E-05
8192	4.99999164	0.83558261E-05
16384	4.99998331	0.16693025E-04

---

[1] These results were obtained on an IBM 1130 computer in so-called "extended precision" which, on that machine, is somewhat less than double-precision.

This time a minimum error is reached at about 2048 trapezoids, and this minimum is significantly lower than the error using standard single-precision arithmetic.

To understand the meaning of roundoff error in this calculation let us plot the error against the number of trapezoids as in Figure 6-5, for both standard and extended precision. In each case the error for both precisions is the same for small numbers of trapezoids, and would in fact decrease linearly (on the logarithmic plot of Figure 6-5) with additional trapezoids were it not for

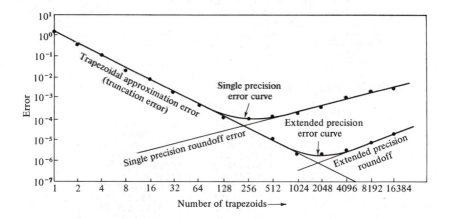

Figure 6-5.   Error curve for Example 6-1.

roundoff error. The standard precision roundoff error is larger, and therefore starts increasing the error near 256 trapezoids, whereas the extended precision calculations can use more trapezoids before roundoff error becomes a problem; hence the minimum is lower and farther to the right.

6.4 Errors and Difficulties of the Trapezoidal Method

Since the integration of Example 6-1 is easily done by hand, we had the advantage of knowing beforehand what the correct answer should be. Hence it is easy to calculate the error and determine the "best" number of trapezoids to use for a given precision. Generally, though, we do not have the benefit of prior knowledge of the answer, and so we have the problem of deciding when to stop taking more trapezoids.

An often used method is to calculate the integral several times, each time with more trapezoids, and stop when two successive answers are almost the same. In Example 6-1 this procedure might have worked to find the minimum error, since the trapezoidal approximation approached the correct integral monotonically, and more important, the roundoff error had the same sign as the error due to taking too few trapezoids (in a manner of speaking, we could consider this error as a truncation error).

Often, however, the roundoff error and the truncation error have opposite signs. As a result, as we take more trapezoids the answer approaches the correct integral until roundoff error becomes large. Then the computed integral crosses to the other side of the correct answer, and the error keeps increasing in the opposite direction. Since there is never any clearcut minimum, it is hard to find the "best" number of trapezoids without having the answer first.

A better way of deciding when to stop is by doing the same calculation with both standard and double-precision, and comparing the two answers. As long as the two answers are approximately the same roundoff error is no problem in either method. When the two answers start to differ by a significant amount, however, this is an indication that standard precision roundoff is becoming large. At this point our past experience leads us to the conclusion that increasing the number of trapezoids by about a factor of 10 will bring us to the point where the extended precision roundoff error will also start to reduce the accuracy; hence we have an idea how many trapezoids to take. This trick will also be useful in the next integration method we shall develop.

As discussed earlier, the trapezoidal rule would provide perfect accuracy for a straight line, and its accuracy decreases as the function to be integrated departs from being straight.

Neglecting roundoff error, it is interesting to note what kind of accuracy we may expect for a given function; that is, the truncation error.

Let us consider the ith trapezoid of a trapezoidal integration, which lies between x_{i-1} and x_i, two points which are a distance $w = (b - a)/n$ apart. Let us further assume that $F(x)$ is the indefinite integral of the function $f(x)$ which we are trying to integrate.

Then the exact value of the integral from x_{i-1} to x_i is

$$I_i = \int_{x_{i-1}}^{x_i} f(x)\, dx = F(x_i) - F(x_{i-1}).$$

The calculated value using the trapezoidal rule, on the other hand, is

$$A_i = \frac{w}{2}[f(x_{i-1}) + f(x_i)].$$

We may now define the error in using the trapezoidal rule on this *one particular trapezoid* as

$$E_i = A_i - I_i. \tag{6-2}$$

We now use Taylor series to simplify. First, we expand $f(x)$ about $x = x_i$ to get $f(x_{i-1})$ in terms of the function and derivatives at $x = x_i$:

$$f(x_{i-1}) = f(x_i) + f'(x_i)(x_{i-1} - x_i) + f''(x_i)\frac{(x_{i-1} - x_i)^2}{2!} + \cdots.$$

But

$$w = x_i - x_{i-1}$$
$$-w = x_{i-1} - x_i,$$

so that

$$f(x_{i-1}) = f(x_i) + f'(x_i)(-w) + f''(x_i)\frac{(-w)^2}{2!} + \cdots$$

$$= f(x_i) - f'(x_i)w + f''(x_i)\frac{w^2}{2!} + \cdots.$$

Next, we have

$$A_i = \frac{w}{2}[f(x_{i-1}) + f(x_i)]$$

$$= \frac{w}{2}\left[f(x_i) + f(x_i) - f'(x_i)w + f''(x_i)\frac{w^2}{2!} + \cdots \right]$$

$$A_i = f(x_i)w - f'(x_i)\frac{w^2}{2} + f''(x_i)\frac{w^3}{2 \cdot 2!} + \cdots. \qquad (6\text{-}3)$$

In the same way we expand the function $F(x)$ about $x = x_i$ to express $F(x_{i-1})$ in terms of the function and its derivatives at $x = x_i$:

$$F(x_{i-1}) = F(x_i) + F'(x_i)(x_{i-1} - x_i) + F''(x_i)\frac{(x_{i-1} - x_i)^2}{2!}$$

$$+ F'''(x_i)\frac{(x_{i-1} - x_i)^3}{3!} + \cdots.$$

As before, $(x_{i-1} - x_i) = -w$, and so

$$F(x_{i-1}) = F(x_i) - F'(x_i)w + F''(x_i)\frac{w^2}{2!} - F'''(x_i)\frac{w^3}{3!} + \cdots.$$

Next, we have

$$I_i = F(x_i) - F(x_{i-1})$$

$$= F(x_i) - \left[F(x_i) - F'(x_i)w + F''(x_i)\frac{w^2}{2!} - F'''(x_i)\frac{w^3}{3!} + \cdots \right]$$

$$= F'(x_i)w - F''(x_i)\frac{w^2}{2!} + F'''(x_i)\frac{w^3}{3!} - \cdots .$$

But since $F(x)$ is the integral of $f(x)$, then

$$f(x) = F'(x)$$
$$f'(x) = F''(x)$$
$$f''(x) = F'''(x),$$
$$\text{etc.}$$

and

$$I_i = f(x_i)w - f'(x_i)\frac{w^2}{2!} + f''(x_i)\frac{w^3}{3!} - \cdots . \tag{6-4}$$

We now substitute Eqs. (6-3) and (6-4) into Eq. (6-2) for the error E_i

$$E_i = A_i - I_i$$

$$= \left[f(x_i)w - f'(x_i)\frac{w^2}{2!} + f''(x_i)\frac{w^3}{2 \cdot 2!} + \cdots \right]$$

$$- \left[f(x_i)w - f'(x_i)\frac{w^2}{2!} + f''(x_i)\frac{w^3}{3!} - \cdots \right]$$

The first two terms in each series cancel, and we have:

$$E_i = f''(x_i)\frac{w^3}{4} - f''(x_i)\frac{w^3}{6} + \text{higher terms in } w^4, w^5, \ldots$$

$$= f''(x_i)w^3[\tfrac{1}{4} - \tfrac{1}{6}] + \text{higher terms.}$$

Assuming that w is small, we may assume that the higher terms in w^4, etc., can be neglected, so that

$$E_i \approx \frac{w^3}{12}f''(x_i).$$

As pointed out earlier, this is the error in the *i*th trapezoid alone. If $|f''(x)| \leqslant M$ for all x in the range $a \leqslant x \leqslant b$, then

$$|E_i| \leqslant \frac{w^3}{12}M.$$

And if this error is summed for n trapezoids, then the maximum total error is

$$|E_T| \leqslant n\frac{w^3}{12}M = nw\frac{w^2}{12}M = (b-a)\frac{w^2}{12}M. \tag{6-5}$$

Hence the truncation error in the trapezoidal method may be expected to be proportional to w^2, the square of the width of each trapezoid, or inversely proportional to n^2, the square of the number of trapezoids. In Figure 6-5 we see that doubling the number of trapezoids reduces the truncation error by a factor of approximately 4.

Since the trapezoidal method is not very accurate because roundoff error becomes a problem before we can use enough small trapezoids to reduce truncation error, we would like to develop a more accurate method, one which has a smaller truncation error.

6.5 Simpson's Method

As shown in Figure 6-2, the trapezoidal method tries to simplify the integration by approximating the curve to be integrated by a series of straight line segments. In Simpson's method we try to approximate by a series of parabolic segments, hoping that the parabola will more closely match a given curve $f(x)$ than would the straight line in the trapezoidal method.

Figure 6-6 shows how we construct the parabola. Suppose we wish to integrate the function $f(x)$ between the limits of a and b as shown. We pick a point $c = (a + b)/2$ midway between a and b and construct the points A, B, and C which have the coordinates

$$A: \quad (a, f(a))$$
$$B: \quad (b, f(b))$$
$$C: \quad (c, f(c)).$$

These three points define a unique parabola $y = \alpha x^2 + \beta x + \gamma$ which passes through all three points. We now hope that the area under the parabola is easier to find than the area under the curve $f(x)$, and that the two areas are approximately equal.

At first glance this may not be too easy, but we shall employ a slight trick by referring back to the trapezoidal method. As you remember, using the trapezoidal method to find the area between $x = a$ and $x = c$ we would have used the equation

$$\text{Area} = \frac{w}{2}[f(a) + f(c)],$$

which is of the general form

$$\text{Area} = Pf(a) + Qf(c),$$

where $P = Q = w/2$ and w is the width of the trapezoid, or the distance between successive points in Figure 6-6.

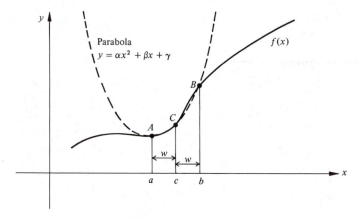

Figure 6-6.   Integration by Simpson's method.

To match the curve by a parabola in Figure 6-6, let us try to derive an equation of a similar form,

$$\text{Area} = Pf(a) + Qf(c) + Rf(b),$$

which should give the area between $x = a$ and $x = b$ if the values of P, Q, and R are properly chosen.

To get P, Q, and R we use a method called the method of *undetermined coefficients*. We note that the trapezoidal method gave us exact answers for any $f(x)$ which was either constant or a straight line since a trapezoid can match these curves exactly. Similarly, Simpson's method should give exact answers for any $f(x)$ which is either constant, or a straight line, or a parabola, since a parabola (possibly an infinitely large one) can match any of these exactly.

To be specific, Simpson's rule should give exact integral answers for the following three integrals:

$$I_1 = \int_{-w}^{+w} 1 \, dx = \left[x \right]_{-w}^{+w} = w - (-w) = 2w$$

$$I_2 = \int_{-w}^{+w} x \, dx = \left[\frac{x^2}{2} \right]_{-w}^{+w} = \frac{w^2}{2} - \frac{(-w)^2}{2} = \frac{w^2}{2} - \frac{w^2}{2} = 0$$

$$I_3 = \int_{-w}^{+w} x^2 \, dx = \left[\frac{x^3}{3} \right]_{-w}^{+w} = \frac{w^3}{3} - \frac{(-w)^3}{3} = \frac{w^3}{3} + \frac{w^3}{3} = \frac{2w^3}{3},$$

which represent the three functions $y = 1$, $y = x$, and $y = x^2$ shown in Figure 6-7.

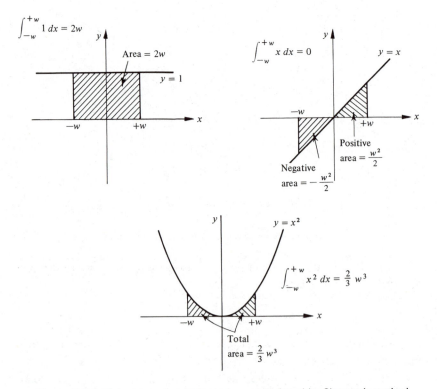

Figure 6-7.   Three integrals which can be exactly found by Simpson's method.

In each case we use $a = -w$, $b = +w$, and the midpoint $c = 0$ and try to get the exact area from an equation of the form

$$\text{Area} = Pf(a) + Qf(c) + Rf(b),$$

so that for the three functions we have

$$
\begin{aligned}
I_1 &= P(1)   &+ Q(1) &+ R(1)   &= P   &+ Q + R &= 2w \\
I_2 &= P(-w)   &+ Q(0) &+ R(w)   &= -Pw   &+ Rw &= 0 \\
I_3 &= P(-w)^2 &+ Q(0)^2 &+ R(+w)^2 = Pw^2 & &+ Rw^2 &= \tfrac{2}{3}w^3 .
\end{aligned}
$$

We therefore have the following three equations in the three unknowns P, Q, and R:

$$
\begin{aligned}
P &+ Q + R &= 2w \\
-Pw &+ Rw &= 0 \\
Pw^2 &+ Rw^2 &= \tfrac{2}{3}w^3 .
\end{aligned}
$$

From the second equation, $Pw = Rw$, so that $P = R$ for any w; inserting $P = R$ into the third equation gives us

$$
Pw^2 + Pw^2 = 2Pw^2 = \tfrac{2}{3}w^3 ,
$$

so that $P = R = w/3$.

Finally, from the first equation,

$$
P + Q + R = 2w
$$
$$
Q = 2w - P - R
$$
$$
= 2w - \frac{w}{3} - \frac{w}{3} = \frac{4}{3}w .
$$

Hence the Simpson's method equation for the area is

$$
\begin{aligned}
\text{Area} &= Pf(a) + Qf(c) + Rf(b) \\
&= \frac{w}{3}f(a) + \frac{4}{3}wf(c) + \frac{w}{3}f(b) \\
&= \frac{w}{3}[f(a) + 4f(c) + f(b)].
\end{aligned}
$$

Since in general our functions will not be straight lines or parabolas, we will find it useful to subdivide the interval to be integrated into an *even* number n of strips of width w, and then use Simpson's rule to find the area of two strips at a time. If, for example, we wish to find the area of the two strips lying between x_{i-1}, x_i, and x_{i+1}, we would use the equation

$$
A_i = \frac{w}{3}[f(x_{i-1}) + 4f(x_i) + f(x_{i+1})]. \tag{6-6}
$$

As in the trapezoidal rule, we would expect the truncation error to decrease as we take more and more (smaller and smaller) intervals; the roundoff error, however, will eventually increase to the point where the overall error increases with smaller intervals. And so we are interested in the speed with which the truncation error drops.

6.6 Truncation Error of Simpson's Method

As an introduction to analyzing the error of Simpson's method, we first try to integrate

$$\int_{-w}^{+w} x^3 \, dx = \left[\frac{x^4}{4}\right]_{-w}^{+w} = \left(\frac{w^4}{4}\right) - \left(\frac{(-w)^4}{4}\right) = \frac{w^4}{4} - \frac{w^4}{4} = 0,$$

with Simpson's method to get

$$\text{Area} = \frac{w}{3}[(-w)^3 + 4(0)^3 + (w)^3] = \frac{w}{3}[-w^3 + w^3] = 0$$

to note that Simpson's method gives the exact answer for this cubic equation; since any cubic function $f(x)$ can be assembled out of linear combinations of x^3, x^2, x, and constants, Simpson's method will be exact for *every* cubic $f(x)$.

To analyze the truncation error, consider the function shown in Figure 6-8, which is subdivided into an even number n of intervals. Taking them two at a time, we find their area by Simpson's rule. Consider the two intervals from x_{i-1} to x_i, and from x_i to x_{i+1} as shown. The error for this set of *two* intervals can be defined as

$$E_i = A_i - I_i, \tag{6-7}$$

where A_i is the area found from Simpson's method

$$A_i = \frac{w}{3}[f(x_{i-1}) + 4f(x_i) + f(x_{i+1})], \tag{6-8}$$

and I_i is the correct area found by integrating the function $f(x)$ from x_{i-1} to x_{i+1}:

$$I_i = \int_{x_{i-1}}^{x_{i+1}} f(x) \, dx = F(x_{i+1}) - F(x_{i-1}). \tag{6-9}$$

We next follow exactly the same procedure as in analyzing the trapezoidal method. First, we simplify Eq. (6-8) with Taylor series by expanding $f(x_{i-1})$ in a series about x_i:

$$f(x_{i-1}) = f(x_i) - f'(x_i)w + f''(x_i)\frac{w^2}{2!} - f'''(x_i)\frac{w^3}{3!} + f^{IV}(x_i)\frac{w^4}{4!} - \cdots,$$

where the minus signs come from the fact $(x_{i-1} - x_i) = -w$ in the Taylor series, and so odd powers of w are negative.

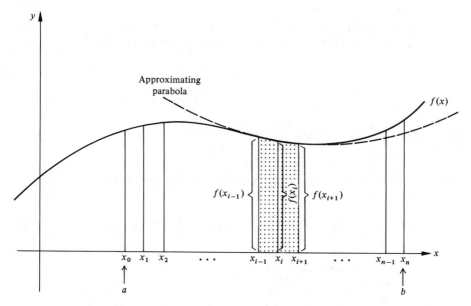

Figure 6-8. Use of Simpson's rule for the two intervals shown.

In exactly the same way we expand $f(x_{i+1})$ about x_i and get exactly the same result except that $(x_{i+1} - x_i) = +w$ and so the odd powers of w are positive:

$$f(x_{i+1}) = f(x_i) + f'(x_i)w + f''(x_i)\frac{w^2}{2!} + f'''(x_i)\frac{w^3}{3!} + f^{IV}(x_i)\frac{w^4}{4!} + \cdots.$$

Inserting into Simpson's rule, Eq. (6-8),

$$A_i = \frac{w}{3}[f(x_{i-1}) + 4f(x_i) + f(x_{i+1})]$$

$$= \frac{w}{3}\left[\left\{f(x_i) - f'(x_i)w + f''(x_i)\frac{w^2}{2!} - f'''(x_i)\frac{w^3}{3!} + f^{IV}(x_i)\frac{w^4}{4!} - \cdots\right\}\right.$$

$$+ 4f(x_i)$$

$$\left. + \left\{f(x_i) + f'(x_i)w + f''(x_i)\frac{w^2}{2!} + f'''(x_i)\frac{w^3}{3!} + f^{IV}(x_i)\frac{w^4}{4!} + \cdots\right\}\right].$$

When we add, all the odd derivatives cancel because of the differing signs, so that

$$A_i = \frac{w}{3}\left[6f(x_i) + 2f''(x_i)\frac{w^2}{2!} + 2f^{IV}(x_i)\frac{w^4}{4!} + \cdots \right]$$

$$= 2f(x_i)w + 2f''(x_i)\frac{w^3}{3!} + 2f^{IV}(x_i)\frac{w^5}{3\cdot 4!} + \cdots. \qquad (6\text{-}10)$$

As the next step, we simplify Eq. (6-9) with Taylor series as we did in the trapezoidal case; first expand $F(x_{i-1})$ in a Taylor series about x_i:

$$F(x_{i-1}) = F(x_i) - F'(x_i)w + F''(x_i)\frac{w^2}{2!} - F'''(x_i)\frac{w^3}{3!} + F^{IV}(x_i)\frac{w^4}{4!}$$

$$- F^V(x_i)\frac{w^5}{5!} + \cdots.$$

In exactly the same way (except for different signs) we have

$$F(x_{i+1}) = F(x_i) + F'(x_i)w + F''(x_i)\frac{w^2}{2!} + F'''(x_i)\frac{w^3}{3!} + F^{IV}(x_i)\frac{w^4}{4!}$$

$$+ F^V(x_i)\frac{w^5}{5!} + \cdots.$$

When we substitute into Eq. (6-9),

$$I_i = F(x_{i+1}) - F(x_{i-1}) \qquad (6\text{-}9)$$

the even terms cancel and odd terms add, and

$$I_i = 2F'(x_i)w + 2F'''(x_i)\frac{w^3}{3!} + 2F^V(x_i)\frac{w^5}{5!} + \cdots.$$

But as before

$$F'(x) = f(x)$$
$$F''(x) = f'(x)$$
$$F'''(x) = f''(x)$$
$$F^V(x) = f^{IV}(x),$$

and therefore

$$I_i = 2f(x_i)w + 2f''(x_i)\frac{w^3}{3!} + 2f^{IV}(x_i)\frac{w^5}{5!} + \cdots. \tag{6-11}$$

We can now calculate the error

$$E_i = A_i - I_i \tag{6-7}$$

by substituting Eqs. (6-10) and (6-11) into (6-7):

$$E_i = \left[2f(x_i)w + 2f''(x_i)\frac{w^3}{3!} + 2f^{IV}(x_i)\frac{w^5}{3\cdot 4!} + \cdots \right]$$

$$- \left[2f(x_i)w + 2f''(x_i)\frac{w^3}{3!} + 2f^{IV}(x_i)\frac{w^5}{5!} + \cdots \right]$$

$$= 2f^{IV}(x_i)\frac{w^5}{3\cdot 4!} - 2f^{IV}(x_i)\frac{w^5}{5!} + \text{higher terms in } w^7 \text{ etc.}$$

Assuming that w is small, we may neglect the terms in w^7 and above, and get the approximate error

$$E_i \approx f^{IV}(x_i)w^5 \left[\frac{2}{3\cdot 4!} - \frac{2}{5!} \right] = f^{IV}(x_i)w^5 \left[\tfrac{2}{72} - \tfrac{2}{120} \right]$$

$$\approx \tfrac{1}{90} f^{IV}(x_i)w^5.$$

If $\left| f^{IV}(x_i) \right| \leqslant M$ for all x in the range of integration, then we may write

$$|E_i| \leqslant \frac{w^5}{90}M.$$

Now if the over-all interval from a to b is divided into n segments, then we repeat the Simpson's method once for each of the $n/2$ pairs of segments, and the total error is

$$|E_S| \leqslant \frac{n}{2}\frac{w^5}{90}M = nw\frac{w^4}{180}M = (b - a)\frac{w^4}{180}M. \tag{6-12}$$

Hence the truncation error of Simpson's method is proportional to w^4, or inversely proportional to n^4, the fourth power of the number of segments used. Hence the error decreases much faster as we take more intervals, and would probably be much lower than the trapezoidal error by the time round-off error starts to cause problems. We can see this in the following example.

6.7 Example 6-2. Integration with Simpson's Method

We shall now use Simpson's method to integrate

$$\int_0^1 (6 - 6x^5)\, dx = 5,$$

which was done in Example 6-1, and is shown in Figure 6-4.

As in the program of Example 6-1, we break up the interval from 0 to 1 into 2^i divisions, where $i = 1, 2, \ldots, 14$; this number is called NODIV in the program and is even. We then divide this number by 2 to give NOSIM, the number of times Simpson's rule is applied (once for each adjacent pair of divisions).

For each iteration we calculate XL and XR, the left and right boundaries, and XM, the center boundary between the two divisions, and the area A of that pair. This area is then added into the total AREA and printed after the end of the loop.

```
      F(X) = 6. - 6. * X ** 5
      WRITE (3, 10)
   10 FORMAT (' DIVISIONS', 6X, 'AREA', 12X, 'ERROR')
      DO 50 I = 1, 14
      NODIV = 2 ** I
      NOSIM = NODIV / 2
      AREA = 0.
      WIDTH = 1. / NODIV
      DO 40 J = 1, NOSIM
      XL =      2. * (J - 1) * WIDTH
      XR =      2. * J * WIDTH
      XM = (XL + XR) / 2.
      A = (WIDTH / 3.) * (F(XL) + 4. * F(XM) + F(XR))
   40 AREA = AREA + A
      ERROR = 5. - AREA
   50 WRITE (3, 60) NODIV, AREA, ERROR
   60 FORMAT (I8, F16.8, E19.8)
      CALL EXIT
      END
```

For comparison we first perform this program in standard (single) precision:

DIVISIONS	AREA	ERROR
2	4.87500001	0.12500098E 00
4	4.99218751	0.78134555E-02
8	4.99951077	0.49018871E-03
16	4.99996758	0.33378608E-04
32	4.99999333	0.76293963E-05
64	4.99999238	0.85830706E-05
128	4.99998284	0.18119815E-04
256	4.99996663	0.34332282E-04
512	4.99993420	0.66757216E-04
1024	4.99985982	0.14114382E-03
2048	4.99971677	0.28419500E-03
4096	4.99947358	0.52738201E-03
8192	4.99892998	0.10709764E-02
16384	4.99783230	0.21686558E-02

and then in double (extended) precision:

DIVISIONS	AREA	ERROR
2	4.87500000	0.12499999E 00
4	4.99218750	0.78125000E-02
8	4.99951172	0.48828125E-03
16	4.99996948	0.30517578E-04
32	4.99999809	0.19073486E-05
64	4.99999985	0.14901161E-06
128	4.99999993	0.70780515E-07
256	4.99999988	0.11920928E-06
512	4.99999974	0.26449561E-06
1024	4.99999948	0.52154064E-06
2048	4.99999901	0.98720192E-06
4096	4.99999788	0.21159648E-05
8192	4.99999579	0.42095780E-05
16384	4.99999169	0.83111226E-05

Figure 6-9 shows the error curves for both standard and extended precision. Comparing this figure with Figure 6-5 (for the trapezoidal method) we see that the roundoff error asymptotes are almost the same for both processes (although the Simpson method, for the same number of divisions, has

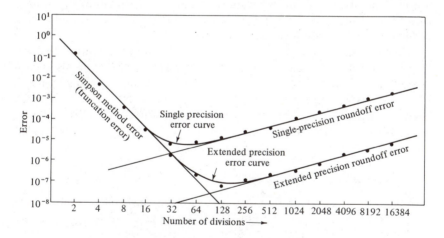

Figure 6-9.   Error curve for Example 6-2.

slightly fewer calculations, and therefore a slightly smaller roundoff error), but the truncation error curve falls much more steeply for Simpson's method than for the trapezoidal method. As a result the minimum errors occur for fewer divisions, and are much lower. Examination of Figure 6-9 shows that doubling the number of divisions decreases the truncation error (neglecting roundoff) by a factor of about $2^4 = 16$, as expected from our previous calculations.

Thus we see that Simpson's method, while slightly more complex as far as the program is concerned, gives us two advantages over the trapezoidal method:

1. Better accuracy is possible at the position of minimum error.
2. Fewer divisions are needed; hence the program runs faster.

6.8 Other Newton-Cotes Methods

The trapezoidal and Simpson's methods are actually two cases of a general series of integration formulas called *Newton-Cotes* methods.

Using the trapezoidal method, we attempted to approximate segments of the function $f(x)$ to be integrated by straight lines which then define trapezoids; in Simpson's method we try to approximate the function $f(x)$ by parabolas. For better accuracy we might try approximation by cubic curves, quartic curves, and so on, and these are all of the Newton-Cotes type.

In the trapezoidal rule, two points (one interval of width w) were used to define a trapezoid; in Simpson's method three points (two intervals of width w each) were used to define a parabola; a cubic approximation would have used four points (three intervals of width w each) to define a cubic curve passing through these four points, and so on.

In each of these Newton-Cotes formulas we have one identical feature; they all use equally-spaced points, which are always the same distance w apart from each other. This leads to easy computation, especially by hand, but not necessarily the best accuracy. In the Gauss method, to be covered shortly, we shall see another approach, the use of unequally-spaced points as an effort to improve the accuracy.

6.9 Romberg Integration

We now go to an integration method which has achieved great popularity in the last few years. To obtain the Romberg method, we shall return temporarily to the trapezoidal method.

In Eq. (6-5) we showed that the total error in doing a trapezoidal integration could be expressed as

$$|E_T| \leqslant (b - a)\frac{w^2}{12}M, \tag{6-5}$$

where w is the width of each trapezoid, and M is the maximum value of the second derivative $f''(x)$ of the function $f(x)$ being integrated. If we let

$$K = \frac{(b - a)}{12}M,$$

then we can write

$$|E_T| \leq Kw^2,$$

so that, neglecting roundoff error, decreasing the size of w makes the truncation error E_T of the trapezoidal method go to zero.

To be specific, for n trapezoids, $w = (b - a)/n$; hence increasing the number of trapezoids by a factor of 2, which reduces w by a factor of 2, should decrease the truncation error by approximately a factor of 4. Figure 6-5 shows that this relation holds in Example 6-1, approximately.

In a way, this is an interesting relation, since it allows us to forecast the error for a trapezoidal integration using more trapezoids without actually performing it. For example, in Figure 6-5 we see that the error for 1024 trapezoids is about 0.3×10^{-5} with extended precision arithmetic. Using 2084 trapezoids should therefore give us an error of about $(0.3 \times 10^{-5})/4$ or about 0.75×10^{-6}, except that roundoff error for this number of trapezoids prevents us from getting a reliable result. In practice, of course, the situation is not as straightforward, since we do not know the correct answer and so cannot know the error. Instead, we proceed as follows: Let

$$I = \text{the exact value of an integral,}$$
$$T_i = \text{the integral evaluated for } i \text{ trapezoids}^2,$$
$$T_{2i} = \text{the integral evaluated for } 2i \text{ trapezoids,}$$

where, for simplicity, we will usually consider i to be some power of 2. Now we could define E_i as the error using i trapezoids, and E_{2i} as the error using $2i$ trapezoids:

$$E_i = T_i - I$$
$$E_{2i} = T_{2i} - I.$$

Assuming that doubling the number of trapezoids decreases the error by a factor of about 4, we have

$$E_{2i} \approx \frac{E_i}{4}$$

$$T_{2i} - I \approx \frac{T_i - I}{4}.$$

---

[2] Don't confuse this T_i with the Chebyshev polynomials $T_n(x)$ defined in Chapter 2; there is no connection.

Simplifying, we have

$$4T_{2i} - 4I \approx T_i - I$$

$$3I \approx 4T_{2i} - T_i$$

$$I \approx \frac{4T_{2i} - T_i}{3} = T_{2i}^{(1)}. \qquad (6\text{-}13)$$

The right-hand side is still only an approximation to I, and we call it $T_{2i}^{(1)}$ to point out that it is the first approximation obtained after calculating T_{2i}, and not the true value.

To point out the utility of this equation, let us take two sets of values obtained (using extended precision) from Example 6-1:

Number of Trapezoids	Trapezoidal Result
$i = 4$	$T_4 = 4.84570312$
$2i = 8$	$T_8 = 4.96105957$

Inserting into Eq. (6-13) we have

$$T_8^{(1)} = \frac{4T_8 - T_4}{3} = 4.99951172,$$

which has an error of approximately 0.49×10^{-3}; the use of the trapezoidal formula alone would require over 64 trapezoids to give comparable accuracy.

As a second example, we start with

Number of Trapezoids	Trapezoidal Result
$i = 128$	$T_{128} = 4.99984729$
$2i = 256$	$T_{256} = 4.99996161$

Inserting into Eq. (6-13) we have

$$T_{256}^{(1)} = \frac{4T_{256} - T_{128}}{3} = 4.99999972.$$

The resulting error is only 0.28×10^{-6}, and is much less than was otherwise possible in Example 6-1 at all.

We now note that the value of $T_8^{(1)}$ is exactly the same as the answer obtained from Simpson's method for 8 intervals (in Example 6-2), and that

$T^{(1)}_{256}$ is remarkably similar to the answer with Simpson's method for 256 intervals. We wonder whether there is any connection, and there is.

Suppose we have three points x_0, x_1, and x_2, separated from each other a distance w, and we wish to integrate a function $f(x)$ between the limits of x_0 and x_2. Using just one trapezoid of width $2w$, we would have

$$T_1 = \frac{2w}{2}[f(x_0) + f(x_2)] = w[f(x_0) + f(x_2)].$$

Using two trapezoids of width w each, on the other hand, we would have

$$T_2 = \underbrace{\frac{w}{2}[f(x_0) + f(x_1)]}_{\substack{\text{first} \\ \text{trapezoid}}} + \underbrace{\frac{w}{2}[f(x_1) + f(x_2)]}_{\substack{\text{second} \\ \text{trapezoid}}}$$

$$= \frac{w}{2}[f(x_0) + 2f(x_1) + f(x_2)].$$

Using Eq. (6-13) we have

$$T^{(1)}_2 = \frac{4T_2 - T_1}{3}$$

$$= \tfrac{1}{3}\{2w[f(x_0) + 2f(x_1) + f(x_2)] - w[f(x_0) + f(x_2)]\}$$

$$= \frac{w}{3}[f(x_0) + 4f(x_1) + f(x_2)],$$

which is just Simpson's formula for the same integral.

So far, it may appear to you that use of Eq. (6-13) is the same as use of Simpson's rule, only in a more complicated way. If we were to stop right here, this would probably seem to be right. We will, however, go a bit further, without actually *proving* the validity of the next step,[3] and extend the method.

Using Eq. (6-13) we have

$$T^{(1)}_{2i} = \frac{4T_{2i} - T_i}{3}. \tag{6-14}$$

In exactly the same way we could have written

$$T^{(1)}_{4i} = \frac{4T_{4i} - T_{2i}}{3}. \tag{6-15}$$

---

[3] The actual proof is rather long, and requires techniques which we shall not cover until later.

These two expressions are successive approximations to the integral, one for $2i$ trapezoids, and the other for $4i$ trapezoids, twice as many. We have already found, however, that $T_{2i}^{(1)}$ is the same as the result obtained from Simpson's method for $2i$ intervals (neglecting roundoff error, of course), while $T_{4i}^{(1)}$ is the same as the result obtained from Simpson's method for $4i$ intervals.

In Eq. (6-12) we showed that the total error in using Simpson's rule can be expressed as

$$|E_S| \leqslant (b - a)\frac{w^4}{180}M. \tag{6-12}$$

The similarity with Eq. (6-5) suggests that we might try to derive a new equation, similar to Eq. (6-13), with which to obtain an even more accurate answer. We let

$I = $ the exact value of the integral

$S_{2i} = T_{2i}^{(1)} = $ the integral evaluated by Simpson's rule using $2i$ intervals, or else obtained from Eq. (6-14)

$S_{4i} = T_{4i}^{(1)} = $ same integral evaluated with $4i$ intervals, or else obtained from Eq. (6-15)

Now we define E_{2i} and E_{4i} as the error using $2i$ and $4i$ intervals, respectively, in Simpson's rule:

$$E_{2i} = S_{2i} - I$$
$$E_{4i} = S_{4i} - I.$$

From Eq. (6-12) we suspect that doubling the number of intervals, from $2i$ to $4i$ reduces the width w of each interval by a factor of 2, and therefore reduces the total error by a factor of $2^4 = 16$. As a rough approximation, we have

$$E_{4i} \approx \frac{E_{2i}}{16}$$

$$S_{4i} - I \approx \frac{S_{2i} - I}{16},$$

and simplifying, we have

$$I \approx \frac{16S_{4i} - S_{2i}}{15} = \frac{16T_{4i}^{(1)} - T_{2i}^{(1)}}{15} = T_{4i}^{(2)}. \tag{6-16}$$

The right-hand side is again only an approximation to I, and we call it $T_{4i}^{(2)}$ to point out that it is the second approximation using the previous $T_{4i}^{(1)}$ and $T_{2i}^{(1)}$.

Following the same steps, we could in fact now use the $T^{(2)}$ approximations to obtain a still better set of approximations $T^{(3)}$, $T^{(4)}$, etc., using the general approximation formula

$$T_{2i}^{(j)} = \frac{4^j T_{2i}^{(j-1)} - T_i^{(j-1)}}{4^j - 1}, \tag{6-17}$$

for $j = 1, 2, \ldots$, and i being any convenient integer (although we are using only powers of two here, this is not necessary).

Using Eq. (6-17) we could obtain an entire series of approximations which, for convenience, might then be expressed in a table like this:

Number of Trapezoids	Trapezoidal Approximation	Further Applications of Eq. (6-17)
1	T_1	
2	$T_2 \longrightarrow T_2^{(1)}$	
4	$T_4 \longrightarrow T_4^{(1)} \to T_4^{(2)}$	
8	$T_8 \longrightarrow T_8^{(1)} \to T_8^{(2)} \to T_8^{(3)}$	
16	$T_{16} \longrightarrow T_{16}^{(1)} \to T_{16}^{(2)} \to T_{16}^{(3)} \to T_{16}^{(4)}$	
32	$T_{32} \longrightarrow T_{32}^{(1)} \to T_{32}^{(2)} \to T_{32}^{(3)} \to T_{32}^{(4)} \to T_{32}^{(5)}$	
64	$T_{64} \longrightarrow T_{64}^{(1)} \to T_{64}^{(2)} \to T_{64}^{(3)} \to T_{64}^{(4)} \to T_{64}^{(5)} \to T_{64}^{(6)}$	
$\vdots$		

The arrows in the table indicate that each term is obtained from the term immediately to its left and the term above that. This process is called *Romberg integration* and, except for roundoff error, should give more accurate results as we proceed to the lower right of the table. As always, we have here a race between truncation error and roundoff error, and we hope that the truncation error will drop to a low enough value before roundoff error becomes a problem.

6.10 Example 6-3. Romberg Integration

As in the previous examples, we again try to integrate

$$\int_0^1 (6 - 6x^5)\, dx = 5,$$

this time using Romberg integration.

The first part of the program, through statement 50, is essentially identical with the trapezoidal program of Example 6-1, except that the T_1 through T_{128} terms are stored in the array $T(1,1)$ through $T(8,1)$. The next part of the program, consisting of the three steps ending at statement 100, calculates the Romberg approximations and stores them in the array T, and the last part of the program prints T in the triangular format.

```
       DIMENSION T(8, 8)
       F(X) = 6. - 6. * X ** 5
C      CALCULATE T-SUB-1 THROUGH T-SUB-128 AND STORE IN ARRAY T(1, 1)
C      THROUGH T(8, 1)
       DO 50 I = 1, 8
       NODIV = 2**(I-1)
       AREA = 0.
       WIDTH = 1. / NODIV
       DO 40 J = 1, NODIV
       XL = (J-1) * WIDTH
       XR = J * WIDTH
       A = (WIDTH / 2.) * (F(XL) + F(XR))
   40  AREA = AREA + A
       T(I, 1) = AREA
   50  CONTINUE
C      NOW CALCULATE THE ROMBERG CORRECTED TERMS
       DO 100 J = 2, 8
       DO 100 I = J, 8
  100  T(I, J) = (4.**(J-1) * T(I, J-1) - T(I-1, J-1)) / (4.**(J-1) - 1)
C      NOW PRINT TABLE OF T'S
       DO 200 I = 1, 8
  200  WRITE (3, 210) (T(I,J), J = 1, I)
  210  FORMAT ('0', 8F11.8)
       CALL EXIT
       END
```

Using extended precision calculations, we obtain the following results from the program:

```
3.00000000

4.40625000 4.87500000

4.84570312 4.99218750 5.00000000

4.96105957 4.99951172 5.00000000 5.00000000

4.99024200 4.99996948 5.00000000 5.00000000 5.00000000

4.99755904 4.99999805 4.99999996 4.99999996 4.99999996 4.99999996

4.99938962 4.99999983 4.99999995 4.99999996 4.99999997 4.99999997 4.99999998

4.99984729 4.99999984 4.99999985 4.99999986 4.99999986 4.99999987 4.99999988 4.99999989
```

The first column consists of T_1 through T_{128}, the same as those obtained with trapezoidal integration (and extended precision) in Example 6-1. The first four terms in the $T^{(1)}$ column match those obtained from Simpson's method in Example 6-2, with following terms slightly different because of different roundoff errors.

The first few terms in the $T^{(2)}$ through $T^{(4)}$ columns provide exact answers while terms lower down and to the right, although theoretically more accurate, exhibit more and more roundoff errors.

We see an interesting fact—$T_4^{(2)}$ is an exact answer. Thus using just four trapezoids, and then applying Eq. (6-17) twice gave better accuracy than thousands of repetitions of Simpson's rule. This helps to explain the popularity of the Romberg method.

6.11 Gauss Quadrature

Gauss investigated the question of whether the use of equally-spaced points, as in Romberg's method or in the Newton-Cotes formulas, limited the accuracy. He found that it did, and developed his own method of numerical integration, called Gauss Quadrature.

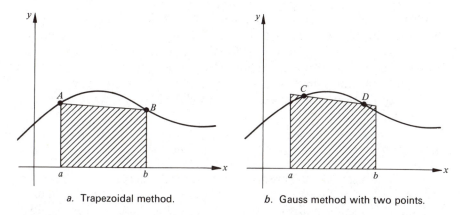

a. Trapezoidal method.   b. Gauss method with two points.

Figure 6-10. Developing the Gauss method for two points from the trapezoidal method.

Figure 6-10 shows a simple curve which we want to integrate between the limits a and b. Part a of Figure 6-10 shows how we would use one trapezoid to do the job: we plot the point A which is at $(a, f(a))$ and the point B at $(b, f(b))$ and draw a straight line connecting the two points. This defines a trapezoid of width $w = (b - a)$, whose area is then

$$A = \frac{w}{2}[f(a) + f(b)].$$

As we pointed out before, this formula is of the form

$$A = Pf(a) + Qf(b),$$

where $P = Q = w/2$.

We now make the approximation that the area under the curve between the limits of a and b is approximately equal to the area of the trapezoid; this is obviously not true, especially not for a large trapezoid as shown in Figure 6-10, and so the trapezoidal method is not very good—we have already found this out.

Using Gauss quadrature, on the other hand, we choose two different points; instead of choosing points A and B at the ends of the interval, we choose two points C and D (as shown in part b of Figure 6-10) which are inside the interval (a, b). We draw a straight line through these points and extend the line out to the ends of the interval, and then complete the shaded trapezoid. Part of the trapezoid lies outside the curve (the upper corners), while part of the curve lies outside the trapezoid. By properly choosing the points C and D, we can balance the two areas so that the area in the trapezoid equals the area under the curve; the resulting approximation then gives us the exact integral. The Gauss method essentially consists of a simple way of choosing C and D to get as good an answer as possible.

With the help of Figure 6-11 let us now show how to go about this. Assume (without loss of generality) that we wish to integrate the function shown in Figure 6-11 between the limits of -1 and $+1$.[4] We choose the points C and D on the curve and complete a trapezoid with the vertices E, F, G, and H.

Suppose that the point C is at $(x_1, f(x_1))$, and the point D is at $(x_2, f(x_2))$. Following the lead of the trapezoidal formula, we would like to develop a formula of the form

$$\text{Area} = C_1 f(x_1) + C_2 f(x_2)$$

since this would make the area relatively simple to calculate. The problem now consists of finding the right values of x_1 and x_2 to use, as well as finding C_1 and C_2 (which would have been equal to $w/2$ in the normal trapezoidal method, but will now be different). There are thus four variables which we can choose for best accuracy.

We shall now adapt the method of undetermined coefficients, which we used in deriving Simpson's method. As you remember, we found the coefficients P, Q, and R by assuming that the method would be exact for functions of the form $y = 1$, $y = x$, and $y = x^2$. This gave us three equations which we solved for the three unknowns P, Q, and R.

This time we have the four unknowns C_1, C_2, x_1, and x_2, and so let us

---

[4] If the limits are different we perform a change of variable to make the limits -1 to $+1$.

assume that the method holds for the four simple functions

$$y = 1$$
$$y = x$$
$$y = x^2$$
$$y = x^3,$$

and solve the four resulting equations for four unknowns. If this works (and it will!) then we have an equation not much harder than the trapezoidal method which will work for cubics, whereas the trapezoidal method worked only for linear curves.

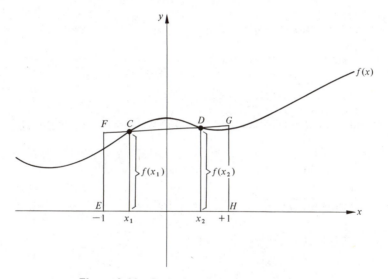

Figure 6-11   Derivation of the Gauss method.

Using the above four simple functions, let us write the following four integrals:

$$I_1 = \int_{-1}^{+1} 1 \, dx = \left[x \right]_{-1}^{+1} = 1 - (-1) = 2$$

$$I_2 = \int_{-1}^{+1} x \, dx = \left[\frac{x^2}{2} \right]_{-1}^{+1} = \frac{1^2}{2} - \frac{(-1)^2}{2} = 0$$

$$I_3 = \int_{-1}^{+1} x^2 \, dx = \left[\frac{x^3}{3} \right]_{-1}^{+1} = \frac{1^3}{3} - \frac{(-1)^3}{3} = \frac{1}{3} - \frac{-1}{3} = \frac{2}{3}$$

$$I_4 = \int_{-1}^{+1} x^3 \, dx = \left[\frac{x^4}{4} \right]_{-1}^{+1} = \frac{1^4}{4} - \frac{(-1)^4}{4} = \frac{1}{4} - \frac{1}{4} = 0.$$

Assuming that an equation of the form

$$\text{Area} = C_1 f(x_1) + C_2 f(x_2)$$

works, then we should have the following four integrals for the four functions above:

$$
\begin{aligned}
I_1 &= C_1(1) \quad + C_2(1) \quad = 2 \\
I_2 &= C_1(x_1) \quad + C_2(x_2) \quad = 0 \\
I_3 &= C_1(x_1)^2 + C_2(x_2)^2 = \tfrac{2}{3}, \\
I_4 &- C_1(x_1)^3 + C_2(x_2)^3 = 0.
\end{aligned}
$$

From the first equation we see that $C_1 + C_2 = 2$; also we note that if

$$C_1 = C_2$$

and

$$x_1 = -x_2,$$

this would satisfy the second and fourth equations. We therefore pick

$$C_1 = C_2 = 1 \quad \text{(so their sum equals 2)}$$
$$x_1 = -x_2$$

and substitute into the third equation as follows:

$$
\begin{aligned}
C_1 x_1^2 + C_2 x_2^2 &= \tfrac{2}{3} \\
x_1^2 + (-x_1)^2 &= \tfrac{2}{3} \\
2x_1^2 &= \tfrac{2}{3} \\
x_1^2 &= \tfrac{1}{3} \\
x_1 &= \pm \frac{1}{\sqrt{3}} = \pm\, 0.57735\ldots
\end{aligned}
$$

We then choose

$$x_1 = -0.57735\ldots$$
$$x_2 = +0.57735\ldots,$$

and obtain the formula

$$
\begin{aligned}
\int_{-1}^{+1} f(x)\, dx &= C_1 f(x_1) + C_2 f(x_2) \\
&= f(-0.57735\ldots) + f(+0.57735\ldots), \qquad \text{(6-18)}
\end{aligned}
$$

which, except for the fact that we have to calculate the value of the function at an irrational value of x, is quite simple.

Before we go to an example, we have two further points. First, the above derivation applies only if we are integrating in the interval from -1 to $+1$. This is no loss in generality, since we can always perform a change of variable to change the interval of integration to be from -1 to $+1$. For example, if we wish to integrate

$$\int_0^\infty e^{-t}\, dt = [-e^{-t}]_0^\infty = -e^{-\infty} + e^0 = 0 + 1 = 1,$$

we could let $t = (x + 1)/(x - 1)$ so that $t = 0$ if $x = -1$, and $t = \infty$ if $x = +1$. Then we find that

$$dt = \frac{(x - 1)\, dx - (x + 1)\, dx}{(x - 1)^2} = -\frac{2\, dx}{(x - 1)^2},$$

so that the integral becomes

$$\int_0^\infty e^{-t}\, dt = \int_{-1}^{+1} \frac{-2e^{-[(x+1)/(x-1)]}}{(x - 1)^2}\, dx,$$

and the conditions of the Gauss method are satisfied.

The other note is the fact that Gauss' method can be extended to three and more points. For example, by choosing three unevenly-spaced points along the curve in the interval -1 to $+1$, we could have passed a parabola through these points, much like Simpson's rule, except that the three points would have been specially chosen to minimize the error. Similarly, we could have used four points and a cubic curve, five points and a quartic, and so on. In each case we would have expected a better fit for higher-order functions.

If we extend the Gauss method to n points, the general equation for the area is

$$A = C_1 f(x_1) + C_2 f(x_2) + C_3 f(x_3) + \cdots + C_n f(x_n).$$

The values of C_i and x_i to be chosen have been tabulated and are easily available; Table 6-1 gives the values for up to 6 points.

The coefficients and abscissas given in Table 6-1 can either be used to integrate over the entire interval of interest, or else the interval can be divided into several smaller subintervals (as in the trapezoidal or Simpson methods) and each of these integrated separately by the Gauss method.

TABLE 6-1

Gauss Quadrature Coefficients and Abscissas

Number of Points (n)	*Coefficients* C_i	*Abscissas* x_i
2	$C_1 = C_2 = 1.0$	$-x_1 = x_2 = 0.57735\,02692$
3	$C_1 = C_3 = 0.555555\ldots$ $C_2 = 0.888888\ldots$	$-x_1 = x_3 = 0.77459\,66692$ $x_2 = 0.0$
4	$C_1 = C_4 = 0.34785\,48451$ $C_2 = C_3 = 0.65214\,51549$	$-x_1 = x_4 = 0.86113\,63116$ $-x_2 = x_3 = 0.33998\,10436$
5	$C_1 = C_5 = 0.23692\,68851$ $C_2 = C_4 = 0.47862\,86705$ $C_3 = 0.568888\ldots$	$-x_1 = x_5 = 0.90617\,98459$ $-x_2 = x_4 = 0.53846\,93101$ $x_3 = 0.0$
6	$C_1 = C_6 = 0.17132\,44924$ $C_2 = C_5 = 0.36076\,15730$ $C_3 = C_4 = 0.46791\,39346$	$-x_1 = x_6 = 0.93246\,95142$ $-x_2 = x_5 = 0.66120\,93865$ $-x_3 = x_4 = 0.23861\,91861$

6.12 Example 6-4. Integration by Gauss Quadrature

We shall now use Gauss quadrature to integrate

$$\int_0^1 (6 - 6x^5)\, dx = 5,$$

which was already done in Examples 6-1 through 6-3. This function is shown in Figure 6-4.

Since this integral has the limits 0 and 1, whereas the Gauss method requires limits of -1 and $+1$, we perform the following change of variable:

$$x = \frac{1 + x'}{2}.$$

Thus when x' is -1, x is 0; when $x' = +1$, $x = 1$, and x' becomes the variable in the Gauss integration. Moreover,

$$dx = \tfrac{1}{2}\, dx',$$

and

$$\int_0^1 (6 - 6x^5) \, dx = \int_{-1}^{+1} \left[6 - 6\left(\frac{1 + x'}{2}\right)^5 \right] (\tfrac{1}{2} \, dx'),$$

and we therefore integrate the second integral. Dropping the primes and re-arranging slightly, we have the new integral

$$\int_{-1}^{+1} (0.5)\left[6 - 6\left(\frac{1 + x}{2}\right)^5 \right] dx,$$

which has the required limits of -1 and $+1$.

The following program defines the integrand with the function statement

FUNC(X) = (0.5) * (6. − 6. * ((1. + X) / 2.) ** 5)

and then uses 2, 3, 4, 5, and 6 points to find the integral by Gauss quadrature.

```
      FUNC(X) = (0.5) * (6. - 6. * ((1. + X) / 2.) ** 5)
      I = 2
      AREA =  FUNC (-0.57735027) +  FUNC (+0.57735027)
      ERROR = 5. - AREA
      WRITE (3, 120) I, AREA, ERROR
  120 FORMAT (' WITH', I3, ' POINTS, THE AREA IS ', F11.7, ' ERROR IS',
     1     E17.8)
      I = 3
      AREA = 0.555555 *  FUNC (-.77459667) + 0.8888889 *  FUNC (0.0)
     1   + 0.5555555 *  FUNC (+.77459667)
      ERROR = 5. - AREA
      WRITE (3, 120) I, AREA, ERROR
      I = 4
      AREA = 0.34785485 *  FUNC (-.86113631)
     1      + 0.65214515 *  FUNC (-.33998104)
     2      + 0.65214515 *  FUNC (+.33998104)
     3      + 0.34785485 *  FUNC (+.86113631)
      ERROR = 5. - AREA
      WRITE (3, 120) I, AREA, ERROR
      I = 5
      AREA = 0.23692689 *  FUNC (-.90617985)
     1      + 0.47862867 *  FUNC (-.53846931)
     2      + 0.56888889 *  FUNC (0.0)
     3      + 0.47862867 *  FUNC (+.53846931)
     4      + 0.23692689 *  FUNC (+.90617985)
      ERROR = 5. - AREA
      WRITE (3, 120) I, AREA, ERROR
      I = 6
      AREA = 0.17132449 *  FUNC (-.93246951)
     1      + 0.36076157 *  FUNC (-.66120939)
     2      + 0.46791393 *  FUNC (-.23861919)
     3      + 0.46791393 *  FUNC (+.23861919)
     4      + 0.36076157 *  FUNC (+.66120939)
     5      + 0.17132449 *  FUNC (+.93246951)
      ERROR = 5. - AREA
      WRITE (3, 120) I, AREA, ERROR
      CALL EXIT
      END
```

The results of this program are

```
WITH  2 POINTS, THE AREA IS   5.0833333 ERROR IS  -0.83333335E-01
WITH  3 POINTS, THE AREA IS   4.9999982 ERROR IS   0.17136335E-05
WITH  4 POINTS, THE AREA IS   5.0000000 ERROR IS   0.37252903E-08
WITH  5 POINTS, THE AREA IS   5.0000000 ERROR IS  -0.74505806E-08
WITH  6 POINTS, THE AREA IS   4.9999999 ERROR IS   0.55879354E-07
```

The Gauss quadrature coefficients and abscissas derived earlier for two points were assumed to hold for functions having up to x^3 powers; three points would work for powers up to x^4; four points would work up to x^5, and so on. Since the integrand is a function with an x^5 power, we would expect that the Gauss quadrature method with 2 or 3 points would give a slight error, while 4 or more points would give an exact answer. This is generally borne out by our results, except that 4 or more points still are affected by roundoff error. Since we only specified the coefficients and abscissas in the program to an accuracy of 8 digits, we certainly cannot expect an answer of better than 8 digit accuracy; hence an error of 10^{-7} or 10^{-8} is about as good as we deserve.

Although the Gauss quadrature integration is substantially more complicated than either trapezoidal or Simpson integration for manual computation, the computer does not care much about the irrational constants required.

Whereas the hundreds (or even thousands) of repetitions required by the trapezoidal or Simpson's methods require seconds with a moderately fast machine using double-precision floating-point arithmetic (a slow computer might even take several minutes to do a trapezoidal or Simpson integration), a Gauss integration of the same function could be done in a fraction of a second.

6.13 Integration of a Series

The integral solved in Examples 6-1 through 6-4 is really much too simple to be done by computer; it is much easier to do it by hand, and gives an exact answer of 5.0 when done by normal calculus. This is often true of simple integrals.

But often we encounter an integral which is either very difficult to solve by hand because we must use some complex change of variable, or perhaps integrate by parts or use some other tricks, or else we may meet an integral which cannot be integrated by ordinary methods at all. There are many such integrals whose indefinite integral cannot be found by the methods of integral calculus, but whose definite integrals are needed for some specific problem.

One such common integral is

$$\int_{-a}^{+a} \frac{1}{\sqrt{2\pi}} e^{-\frac{1}{2}x^2} \, dx,$$

which describes the familiar "bell curve" (similar to Figure 8-6) occurring in probability and statistics. We cannot find the indefinite integral by any means, and yet often need the definite integral between some limits $-a$ and $+a$.

The digital computer now comes to the rescue by permitting solutions which could not otherwise be obtained. The above integral could be integrated by the trapezoidal, Simpson, Romberg, or Gauss methods, discussed previously, but in this case another method is also possible: integration of a series. But it cannot always be used.

To use this method we express the integrand as an infinite Taylor or Maclaurin series. If the series converges *uniformly* to the sum in the interval of integration, then we can integrate the series term by term, and then sum the integrated terms. This approach becomes practical when we have a digital computer to do the summing, as we shall see in the following example.

6.14 Example 6-5. Integration of a Uniformly Convergent Series

Consider the integral

$$\int_{-1}^{+1} \frac{1}{\sqrt{2\pi}} e^{-\frac{1}{2}x^2} \, dx = \frac{1}{\sqrt{2\pi}} \int_{-1}^{+1} e^{-\frac{1}{2}x^2} \, dx,$$

which cannot be solved by ordinary calculus. Since the integrand is an exponential, we may use a Maclaurin series for e^a as follows. Since

$$e^a = 1 + a + \frac{a^2}{2!} + \frac{a^3}{3!} + \frac{a^4}{4!} + \frac{a^5}{5!} + \frac{a^6}{6!} + \frac{a^7}{7!} + \cdots,$$

we may let $a = -\frac{1}{2}x^2$ and write the series as

$$e^{-\frac{1}{2}x^2} = 1 - \frac{x^2}{2} + \frac{x^4}{2^2 \cdot 2!} - \frac{x^6}{2^3 \cdot 3!} + \frac{x^8}{2^4 \cdot 4!} - \frac{x^{10}}{2^5 \cdot 5!} + \frac{x^{12}}{2^6 \cdot 6!} - \frac{x^{14}}{2^7 \cdot 7!} + \cdots$$

$$= 1 - \frac{x^2}{2} + \frac{x^4}{8} - \frac{x^6}{48} + \frac{x^8}{384} - \frac{x^{10}}{3840} + \frac{x^{12}}{46080} - \frac{x^{14}}{645120} + \cdots.$$

The integral is now

$$\frac{1}{\sqrt{2\pi}} \int_{-1}^{+1} e^{-\frac{1}{2}x^2}\, dx = \frac{1}{\sqrt{2\pi}} \int_{-1}^{+1} \left\{ 1 - \frac{x^2}{2} + \frac{x^4}{8} - \frac{x^6}{48} + \frac{x^8}{384} - \frac{x^{10}}{3840} \right.$$
$$\left. + \frac{x^{12}}{46080} - \frac{x^{14}}{645120} + \cdots \right\}\, dx.$$

Since the series converges uniformly to the exponential in the range $-1 \leqslant x \leqslant +1$ (proof is left to the reader) we can integrate the series term by term, to get

$$\frac{1}{\sqrt{2\pi}} \int_{-1}^{+1} e^{-\frac{1}{2}x^2}\, dx$$

$$= \frac{1}{\sqrt{2\pi}} \left[\int_{-1}^{+1} 1\, dx - \int_{-1}^{+1} \frac{x^2}{2}\, dx + \int_{-1}^{+1} \frac{x^4}{8}\, dx - \int_{-1}^{+1} \frac{x^6}{48}\, dx \right.$$
$$+ \int_{-1}^{+1} \frac{x^8}{384}\, dx - \int_{-1}^{+1} \frac{x^{10}}{3840}\, dx + \int_{-1}^{+1} \frac{x^{12}}{46080}\, dx$$
$$\left. - \int_{-1}^{+1} \frac{x^{14}}{645120}\, dx + \cdots \right].$$

Each of these integrals is easy to integrate, except that there is an infinite number of them! If we start to integrate the first few terms, we get

$$\frac{1}{\sqrt{2\pi}} \int_{-1}^{+1} e^{-\frac{1}{2}x^2}\, dx$$

$$= \frac{1}{\sqrt{2\pi}} \left\{ \left[x \right]_{-1}^{+1} - \left[\frac{x^3}{6} \right]_{-1}^{+1} + \left[\frac{x^5}{40} \right]_{-1}^{+1} - \left[\frac{x^7}{336} \right]_{-1}^{+1} + \left[\frac{x^9}{3456} \right]_{-1}^{+1} \right.$$
$$\left. - \left[\frac{x^{11}}{42240} \right]_{-1}^{+1} + \left[\frac{x^{13}}{599040} \right]_{-1}^{+1} - \left[\frac{x^{15}}{9676800} \right]_{-1}^{+1} + \cdots \right\}.$$

Finally, substituting the limits into each bracket we get

$$\frac{1}{\sqrt{2\pi}} \int_{-1}^{+1} e^{-\frac{1}{2}x^2}\, dx = \frac{2}{\sqrt{2\pi}} \left[1 - \frac{1}{6} + \frac{1}{40} - \frac{1}{336} + \frac{1}{3456} - \frac{1}{42240} \right.$$
$$\left. + \frac{1}{599040} - \frac{1}{9676800} + \cdots \right].$$

With some analysis, we can see that this series can be written in the compact form

$$\frac{1}{\sqrt{2\pi}} \int_{-1}^{+1} e^{-\frac{1}{2}x^2} \, dx = \frac{2}{\sqrt{2\pi}} \sum_{n=0}^{\infty} \frac{(-1)^n}{2^n \cdot n! \cdot (2n+1)}.$$

Since this is an alternating series which obviously converges, the computer can find the sum using the techniques of Chapter 2.

For illustrative purposes, and to permit us to compare the five methods of integration of this chapter, the following program sums the above series for the first eight terms (accurate to at least 10^{-7} since it is an alternating series and the first term neglected is smaller than 10^{-7}), and also performs the integration by each of the other four methods. Since we have definite knowledge of the series truncation error, integration of the series is possibly the safest method to use in this case; unfortunately it often cannot be used.

```
      DIMENSION T(13, 13)
      EFUNC (XI) = EXP(-XI * XI / 2.) / SQRT (2. * PI)
      PI = 3.1415926
C     AREA BY INTEGRATION OF THE SERIES
      WRITE (3, 10)
   10 FORMAT (' INTEGRATION OF THE POWER SERIES', /)
      AREA = 2.*(1. - 1./6. + 1./40. - 1./336. + 1./3456. - 1./42240.
     1    + 1./599040. - 1./9676800.) / SQRT (2. * PI)
      WRITE (3, 20) AREA
   20 FORMAT (20X, F10.7)
C AREA BY THE TRAPEZOIDAL RULE
      WRITE (3, 30)
   30 FORMAT('0INTEGRATION BY TRAPEZOIDAL RULE', /)
      DO 50 I = 1, 13
      NODIV = 2 ** I / 2
      AREA = 0.
      WIDTH = 2. / NODIV
      DO 40 J = 1, NODIV
      XL = -1. + (J-1) * WIDTH
      XR = -1. + J * WIDTH
      A = (WIDTH / 2.) * (EFUNC (XL) + EFUNC (XR))
   40 AREA = AREA + A
      T(I, 1) = AREA
   50 WRITE (3, 60) NODIV, AREA
   60 FORMAT (' WITH', I6, ' TRAPEZOIDS, THE AREA IS ', F11.7)
C     CALCULATE ROMBERG INTEGRATION TERMS THROUGH T-SUB-16, USING THE
C          TRAPEZOIDAL AREAS STORED IN THE ARRAY T
      WRITE (3, 61)
   61 FORMAT ('0ROMBERG INTEGRATION', /)
      DO 62 J = 2, 5
      DO 62 I = J, 5
   62 T(I, J) = (4.**(J-1) * T(I,J-1) - T(I-1,J-1)) / (4.**(J-1) - 1)
      DO 65 I = 1, 5
   65 WRITE (3, 67) (T(I, J), J = 1, I)
   67 FORMAT (5F12.7)
C AREA BY SIMPSON'S RULE
      WRITE (3, 70)
   70 FORMAT ('0INTEGRATION BY SIMPSON''S RULE', /)
      DO 90 I = 1, 12
      NODIV = 2 ** I
```

```
       NOSIM = NODIV / 2
       AREA = 0.
       WIDTH = 2. / NODIV
       DO 80 J = 1, NOSIM
       XL = -1. + (J-1) * 2. * WIDTH
       XR = -1. + J * 2. * WIDTH
       XM = (XL + XR) / 2.
       A = (WIDTH/3.) * (EFUNC (XL) + 4. * EFUNC (XM) + EFUNC (XR))
    80 AREA = AREA + A
    90 WRITE (3, 100) NODIV, AREA
   100 FORMAT (' WITH', I6, ' DIVISIONS, THE AREA IS ', F11.7)
C AREA BY GAUSS QUADRATURE
       WRITE (3, 110)
   110 FORMAT ('0INTEGRATION BY GAUSS QUADRATURE', /)
       I = 2
       AREA = EFUNC (-0.57735027) + EFUNC (+0.57735027)
       WRITE (3, 120) I, AREA
   120 FORMAT (' WITH', I3, ' POINTS, THE AREA IS ', F11.7)
       I = 3
       AREA = 0.555555 * EFUNC (-.77459667) + 0.8888889 * EFUNC (0.0)
      1  + 0.5555555 * EFUNC (+.77459667)
       WRITE (3, 120) I, AREA
       I = 4
       AREA = 0.34785485 * EFUNC (-.86113631)
      1      + 0.65214515 * EFUNC (-.33998104)
      2      + 0.65214515 * EFUNC (+.33998104)
      3      + 0.34785485 * EFUNC (+.86113631)
       WRITE (3, 120) I, AREA
       I = 5
       AREA = 0.23692689 * EFUNC (-.90617985)
      1      + 0.47862867 * EFUNC (-.53846931)
      2      + 0.56888889 * EFUNC (0.0)
      3      + 0.47862867 * EFUNC (+.53846931)
      4      + 0.23692689 * EFUNC (+.90617985)
       WRITE (3, 120) I, AREA
       I = 6
       AREA = 0.17132449 * EFUNC (-.93246951)
      1      + 0.36076157 * EFUNC (-.66120939)
      2      + 0.46791393 * EFUNC (-.23861919)
      3      + 0.46791393 * EFUNC (+.23861919)
      4      + 0.36076157 * EFUNC (+.66120939)
      5      + 0.17132449 * EFUNC (+.93246951)
       WRITE (3, 120) I, AREA
       CALL EXIT
       END
```

Using extended precision, the program gives us the following results:

```
INTEGRATION OF THE POWER SERIES

                0.6826894

INTEGRATION BY TRAPEZOIDAL RULE

   WITH      1 TRAPEZOIDS, THE AREA IS   0.4839414
   WITH      2 TRAPEZOIDS, THE AREA IS   0.6409130
   WITH      4 TRAPEZOIDS, THE AREA IS   0.6725218
   WITH      8 TRAPEZOIDS, THE AREA IS   0.6801636
   WITH     16 TRAPEZOIDS, THE AREA IS   0.6820590
   WITH     32 TRAPEZOIDS, THE AREA IS   0.6825319
   WITH     64 TRAPEZOIDS, THE AREA IS   0.6826501
   WITH    128 TRAPEZOIDS, THE AREA IS   0.6826796
   WITH    256 TRAPEZOIDS, THE AREA IS   0.6826870
   WITH    512 TRAPEZOIDS, THE AREA IS   0.6826888
```

```
WITH  1024 TRAPEZOIDS, THE AREA IS    0.6826892
WITH  2048 TRAPEZOIDS, THE AREA IS    0.6826891
WITH  4096 TRAPEZOIDS, THE AREA IS    0.6826889
```

```
ROMBERG INTEGRATION

   0.4839414
   0.6409130    0.6932368
   0.6725218    0.6830581    0.6823795
   0.6801636    0.6827109    0.6826878    0.6826927
   0.6820590    0.6826908    0.6826894    0.6826894    0.6826894
```

```
INTEGRATION BY SIMPSON'S RULE

WITH     2 DIVISIONS, THE AREA IS    0.6932368
WITH     4 DIVISIONS, THE AREA IS    0.6830581
WITH     8 DIVISIONS, THE AREA IS    0.6827109
WITH    16 DIVISIONS, THE AREA IS    0.6826908
WITH    32 DIVISIONS, THE AREA IS    0.6826895
WITH    64 DIVISIONS, THE AREA IS    0.6826895
WITH   128 DIVISIONS, THE AREA IS    0.6826894
WITH   256 DIVISIONS, THE AREA IS    0.6826894
WITH   512 DIVISIONS, THE AREA IS    0.6826894
WITH  1024 DIVISIONS, THE AREA IS    0.6826894
WITH  2048 DIVISIONS, THE AREA IS    0.6826893
WITH  4096 DIVISIONS, THE AREA IS    0.6826892
```

```
INTEGRATION BY GAUSS QUADRATURE

WITH  2 POINTS, THE AREA IS    0.6753947
WITH  3 POINTS, THE AREA IS    0.6829970
WITH  4 POINTS, THE AREA IS    0.6826798
WITH  5 POINTS, THE AREA IS    0.6826897
WITH  6 POINTS, THE AREA IS    0.6826894
```

A comparison of results is interesting. Of all the methods, the fastest (in computer time) and most reliable is the summation of the series, since it is an alternating series and we know what its maximum error will be; for least roundoff error, we should probably sum the series backward, however. The trapezoidal rule gives a very close answer with 1024 trapezoids, but roundoff error prevents getting any closer. Romberg integration provides the right answer to seven decimal places on the last line, which represents $T_{16}^{(2)}$ through $T_{16}^{(4)}$. Here, only 16 trapezoids followed by repeated application of the Romberg method is sufficient to give us the right answer, at very little work.

Simpson's method provides 7-place accuracy if we pick the right number of divisions, but gives a slight error if we pick too many. Gauss quadrature gives the exact answer with only 6 points and probably represents the least amount of actual computation; the trouble with it is that the error analysis is difficult, and it is difficult to recognize the right answer when we have it.

6.15 Problems

1. The programs of Examples 6-1 and 6-2 can be shortened somewhat by noting that, for each iteration after the first, *XL* and *F(XL)* are actually equal to XR and $F(XR)$ from the previous iteration, and thus need not be calculated again. Rewrite the programs to take advantage of this fact.

2. Proceeding in the same way as in deriving Simpson's rule from the trape-zoidal rule and Eq. (6-13), use Simpson's rule and Eq. (6-17) as a starting point to derive the fourth-order Newton-Cotes formula, also known as Boole's rule:

$$T_4^{(2)} = \frac{2w}{45}[7f(x_0) + 32f(x_1) + 12f(x_2) + 32f(x_3) + 7f(x_4)].$$

3. Figure 6-12 shows the graph of the function $f(x) = \sin x/x$. Consider carefully the problems in integrating

$$\int_0^\pi \frac{\sin x}{x}\, dx$$

by each of the methods of this chapter, especially with regard to the point

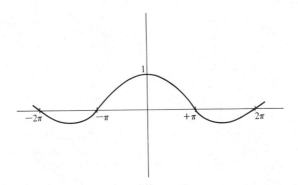

Figure 6-12.   Function sin x/x for problem 3.

at $x = 0$. Prepare programs using those methods which you consider to be "safe."

4. What criteria might we use to decide when to stop a Romberg integration program, and which of its results to consider the final answer?

5. Rather than apply Simpson's rule individually to each pair of divisions, show that the interval to be integrated could be first divided into n divisions (n being even), and then the integral evaluated with the equation

$$\frac{w}{3}[f(x_0) + 4f(x_1) + 2f(x_2) + 4f(x_3) + \cdots + 2f(x_{n-2})$$
$$+ 4f(x_{n-1}) + f(x_n)].$$

6. Every calculus text has a table of common integrals such as the following; it is sometimes interesting to check this table for some definite integrals numerically. For the following, pick appropriate limits of integration (making sure to avoid values of x for which the integrand becomes infinite) and perform a numerical integration using the five methods of this chapter; also evaluate the right-hand side of each expression at the limits to check your answer.

a. $\displaystyle\int e^x \, dx = e^x + C;$

b. $\displaystyle\int b^x \, dx = \frac{b^x}{\ln b} + C$ [choose $b > 0$ but $b \neq 1$];

c. $\displaystyle\int \sin x \, dx = -\cos x + C;$

d. $\displaystyle\int \cos x \, dx = \sin x + C;$

e. $\displaystyle\int \tan x \, dx = -\ln \cos x + C;$

f. $\displaystyle\int \operatorname{ctn} x \, dx = \ln \sin x + C;$

g. $\displaystyle\int \sec x \, dx = \ln \tan \left(\frac{x}{2} + \frac{\pi}{4}\right) + C;$

h. $\displaystyle\int \csc x \, dx = \ln \tan \frac{x}{2} + C;$

i. $\displaystyle\int \sin^2 x \, dx = \tfrac{1}{2}x - \tfrac{1}{2}\sin x \cos x + C;$

j. $\displaystyle\int \cos^2 x \, dx = \tfrac{1}{2}x + \tfrac{1}{2}\sin x \cos x + C;$

k. $\displaystyle\int \tan^2 x \, dx = \tan x - x + C;$

l. $\displaystyle\int \frac{dx}{x^2 - a^2} = \frac{1}{2a} \ln \left(\frac{x - a}{x + a}\right) + C$ [choose $x^2 > a^2$];

m. $\displaystyle\int \frac{dx}{x^2 + a^2} = \frac{1}{a} \tan^{-1} \frac{x}{a} + C;$

n. $\displaystyle\int \frac{1}{x} dx = \ln x + C.$

7. Try various methods of integration on

$$\int_{-1}^{+1} \frac{x^2 \, dx}{\sqrt{1 - x^2}}$$

What difficulties do you encounter, and how might you overcome them?

7

Ordinary Differential Equations

In this chapter we will examine computer solution of ordinary differential equations from a beginner's viewpoint.

Consider the following equation:

$$y = x^2.$$

This equation is *not* a differential equation; rather, it is the now familiar equation of a parabola in the x–y plane. But suppose we differentiate the equation with respect to x to get

$$\frac{dy}{dx} = 2x.$$

The new equation, in its own way, also describes a parabola. But instead of explicitly giving a value of y for every value of x, the new equation describes the *slope* or *direction* of the parabola at every point.

The equation $dy/dx = 2x$ is called an *ordinary differential equation*; differential because it is a relation between derivatives, and ordinary because the derivative happens to be the ordinary derivative dy/dx (as opposed to a partial derivative $\partial y/\partial x$).

235

Furthermore, since the original equation $y = x^2$ satisfies the requirement that $dy/dx = 2x$ for every x, we say that $y = x^2$ is a *solution* of the differential equation $dy/dx = 2x$.

7.1 Introduction to Differential Equations

A differential equation may be described as an equation containing derivatives, such as the following:

$$(1) \quad \frac{dy}{dx} = 2x; \qquad\qquad (3) \quad \frac{d^2y}{dx^2} + 3\frac{dy}{dx} - 17y = 0;$$

$$(2) \quad \frac{dy}{dx} = x^2 + y^2; \qquad (4) \quad \left(\frac{d^3y}{dx^3}\right)^2 - 15\frac{d^2y}{dx^2} + 2y = 15.$$

For the sake of completeness, we should define a few of the standard terms for differential equations:

Definition: *The* dependent variable *is the variable being differentiated.* (*In each of the above examples,* y *is the dependent variable.*)

Definition: *The* independent variable *is the variable with respect to which the differentiations are performed.* (*In each of the above examples,* x *is the independent variable.*)

Definition *The* order *of a differential equation is the order of the highest derivative.*

Definition *The* degree *of a differential equation is the power to which the highest-order derivative is raised.*

Definition *A differential equation is called* homogeneous *if every term of the equation contains the dependent variable; otherwise, it is called* nonhomogeneous.

As an example, the two equations

$$\frac{dy}{dx} = 2x \quad \text{and} \quad \frac{dy}{dx} = x^2 + y^2$$

are both first-order, first-degree nonhomogeneous equations;

$$\frac{d^2y}{dx^2} + 3\frac{dy}{dx} - 17y = 0$$

is a homogeneous differential equation of first degree, second order; and

$$\left(\frac{d^3y}{dx^3}\right)^2 - 15\frac{d^2y}{dx^2} + 2y = 15$$

is a nonhomogeneous equation of second degree and third order.

In this chapter we shall consider the solution of first-order, first-degree equations of the type

$$\frac{dy}{dx} = f(x, y),$$

where $f(x, y)$ can be any single-valued and analytic function of either x or y, or both, such as $2x$, $(x^2 + y^2)$, or even $(3 + x \cos y)$. Since we are placing no other restrictions on $f(x, y)$, we can therefore consider both homogeneous and nonhomogeneous equations.

7.2 Analytic Solutions and Computer Solutions

Starting with the solution, it is usually very easy to construct a differential equation by just differentiating. For example, examine the following list. In each case, we start with a "solution" at the left, differentiate, and finish with the differential equation at the right:

"Solution"	Differential Equation
$y = x^3$	$\dfrac{dy}{dx} = 3x^2$
$y = \sin x$	$\dfrac{dy}{dx} = \cos x$
$y = 15$	$\dfrac{dy}{dx} = 0$
$y = e^x$	$\dfrac{dy}{dx} = y(= e^x)$
$y = e^{2x}$	$\dfrac{dy}{dx} = 2y$
$y = e^{2x} + \sin x$	$\dfrac{dy}{dx} = 2y + \cos x - 2\sin x$

The right-hand side of the above differential equations is what we call

$f(x, y)$, since it may be a function of either x or y or both (or neither, as in a constant).

But the above process is somewhat like the tail wagging the dog; we do not usually start with the "solution" and then manufacture a differential equation out of it. Instead, we often have a differential equation for which we need the solution. Moreover, finding the solution is not always easy and often requires quite a bit of ingenuity.

Thus, when a mathematician says that he is "solving" a differential equation of the general form $dy/dx = f(x, y)$, he means that he is seeking an equation of the form $y = F(x)$ which, if differentiated or otherwise manipulated, will give the original differential equation. Once he has the relation $y = F(x)$, he can start with any value of x and find the corresponding value of y. Such a $y = F(x)$ is called an *analytic* solution.

(Just to emphasize the fact that this is not easy, examine the differential equation

$$\frac{dy}{dx} = y + e^x.$$

It is not at all obvious that $y = (x - 1) e^x$ is a solution.)

Unfortunately, using computer methods and numerical analysis, we are unable to find an *analytic* solution such as $y = F(x)$ since a computer must work with numbers, not with functions. Fortunately, we really do not need to do this, since the usual purpose of finding a solution is to be able to find the value of y which corresponds to any given x. Using a *computer* solution we can do this without actually finding the function $F(x)$.

7.3 Initial or Boundary Conditions

Unfortunately, the problem of finding *a* solution is complicated by the fact that there is an *infinite* number of them!

Let us return to the differential equation we had at the start of this chapter,

$$\frac{dy}{dx} = 2x.$$

The right-hand side is our $f(x, y)$, and since it equals a derivative dy/dx, can be thought of as a direction which depends on the x coordinate. In fact, Figure 7-1 shows that we can interpret $dy/dx = 2x$ as a set of direction vectors in the x–y plane, so that associated with any point (x, y) there is a specific direction vector.

If $f(x, y)$ happens to be a function of x alone, as in our example, then for a given x the direction is the same regardless of y. Thus, at every point on the

y axis, where $x = 0$, the direction vectors are horizontal; at $x = 1$ the direction vectors have a slope of $+2$, and so on.

If, on the other hand, $f(x, y)$ were a function of y alone, then the direction vector would be independent of x, whiie in the general case, the direction vector will depend on both x and y.

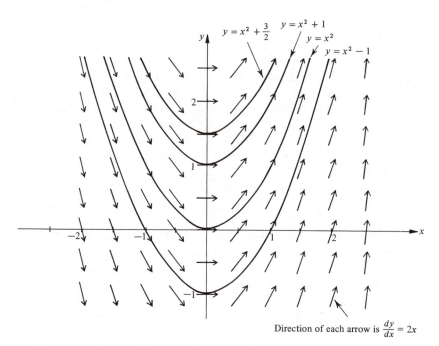

Direction of each arrow is $\dfrac{dy}{dx} = 2x$

Figure 7-1.   Direction field and several solutions of the equation $dy/dx = 2x$.

Any solution to the differential equation must lie in the direction of the field at every point. Unfortunately, we can see from Figure 7-1 that there are any number of curves which satisfy the direction field. Since the direction field of our example is independent of the y coordinate, we could take any solution curve and move it any arbitrary distance in the y direction. Hence any curve of the form

$$y = x^2 + c$$

will satisfy the differential equation (and its direction field) regardless of the value of c.

We could have come to the same conclusion mathematically. Starting with the differential equation

$$\frac{dy}{dx} = 2x,$$

we rearrange and integrate:

$$dy = 2x \, dx$$

$$\int dy = \int 2x \, dx$$

$$y = x^2 + c,$$

where c is a constant of integration which can take on any value. Thus there is an infinite number of solutions.

In order to pick out one specific solution out of the infinite number, we need more information than is contained in the differential equation alone. This is given in the form of an *initial condition* (which, in more complicated equations, is also often called a *boundary condition* or *boundary value*). To pick one specific curve in Figure 7-1, we could specify that the solution, besides satisfying the differential equation, must also pass through some specified point, say (x_0, y_0).

This raises a question: does providing a specific point pick out a specific solution? In other words, could *two* solution curves pass through this *one* point?

Let us temporarily assume that this is possible. Then, for both solutions to pass through the same point, they must either (1) cross over at that point, or (2) be tangent to each other.

The first case is obviously impossible, since if the two curves were to cross over they would have different derivatives $dy/dx = f(x_0, y_0)$. But if $f(x, y)$ is a single-valued function then only one direction is possible at a given point (x_0, y_0).

The second case is harder to eliminate. Suppose that the two solutions are functions of the forms

$$y = F_1(x), \qquad \text{such that } y_0 = F_1(x_0)$$

and

$$y = F_2(x), \qquad \text{such that } y_0 = F_2(x_0).$$

Let us expand $F_1(x)$ and $F_2(x)$ in Taylor series about the point $x = x_0$:

$$F_1(x) = F_1(x_0) + F_1'(x_0)(x - x_0) + F_1''(x_0)\frac{(x - x_0)^2}{2!} + \cdots$$

$$F_2(x) = F_2(x_0) + F_2'(x_0)(x - x_0) + F_2''(x_0)\frac{(x - x_0)^2}{2!} + \cdots.$$

We now have $F_1(x_0) = F_2(x_0) = y_0$, as well as the fact that $F_1'(x_0) = F_2'(x_0) = f(x_0, y_0)$ since both functions are tangent to each other and have the same derivative at (x_0, y_0).

So we have:

$$F_1(x) = y_0 + f(x_0, y_0)(x - x_0) + F_1''(x_0)\frac{(x - x_0)^2}{2!} + \cdots$$

$$F_2(x) = y_0 + f(x_0, y_0)(x - x_0) + F_2''(x_0)\frac{(x - x_0)^2}{2!} + \cdots.$$

For $F_1(x)$ and $F_2(x)$ to be different, there must be some region where they have different values; in this region their Taylor series must sum to different values, and so the Taylor series must be different. This means that at least one set of higher derivatives must be different. But $F_1''(x)$ and $F_2''(x)$ can both be obtained by differentiating $f(x, y)$ with respect to x, and so they cannot differ. Similarly, $F_1'''(x)$ and $F_2'''(x)$ are obtained by a double differentiation and so they too must be the same, and so on. Hence $F_1(x)$ and $F_2(x)$ cannot be different.

And so we find that providing the additional condition that the solution must pass through a given point is enough to specify a unique solution.

Let us return to our example of $dy/dx = 2x$, and add the condition that the solution must pass through the point $(1, 2)$. We can express this condition in a slightly different way by noting that each of the curves in Figure 7-1 represents an analytic equation of the form

$$y = F(x) = x^2 + c.$$

We might use this analytic notation to say that

$$F(1) = 2.$$

As before, this new notation says that y should equal 2 when x equals 1. We can then substitute into the equation for y to find the constant c as follows:

$$F(x) = x^2 + c$$
$$F(1) = 1^2 + c = 2$$
$$1 + c = 2$$
$$c = 2 - 1 = 1.$$

Then the unique solution must be

$$F(x) = x^2 + 1,$$

and we have found the answer.

As a further example of how initial conditions affect analytic solutions, consider the differential equation

$$\frac{dy}{dx} = y.$$

This happens to be a relatively simple equation which we can solve by re-arranging and integrating:

$$\frac{dy}{dx} = y$$

$$\frac{dy}{y} = dx$$

$$\int \frac{dy}{y} = \int dx$$

$$\ln y = x + c.$$

Taking exponentials of both sides, we have

$$e^{\ln y} = y = e^{x+c} = e^x e^c,$$

and if we let $k = e^c$ (which is possible since c is a constant and so is k), we finally have

$$y = k e^x.$$

Since c could be any constant of integration, $k = e^c$ can also be any positive constant. In fact, we can show that $y = k e^x$ is a solution of the differential equation even if k is a negative constant, since if we differentiate

$$y = k e^x$$

$$\frac{dy}{dx} = k e^x,$$

and therefore, for any k,

$$\frac{dy}{dx} = y.$$

The solution again has an unknown constant k, so that there could be an infinite number of solutions. But if we specify an initial condition, either by saying that the curve $y = F(x) = k\,e^x$ must pass through a given point (x_0, y_0) or by simply stating that $F(x_0) = y_0$, then a specific solution is obtained since only one value of k will satisfy the initial condition.

7.4 Euler's Method

The analytic method of solving a differential equation as in the preceding sections is not very useful in solving the problem on a computer. Instead, we rely on numerical methods.

The simplest numerical method for solving a differential equation is called

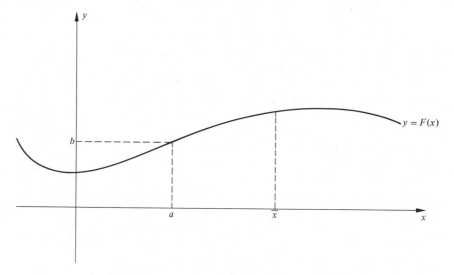

Figure 7-2.   The solution of a differential equation.

Euler's method. Although it is far from being accurate, it serves as a good introduction and so we will look into it.

Let us suppose that we are given a differential equation of the form

$$\frac{dy}{dx} = f(x, y)$$

along with the initial condition that $y = b$ when $x = a$. This initial condition therefore picks one specific solution out of the infinite number. In general, the curve shown in Figure 7-2 might be such a solution, passing through the point (a, b). If we were to solve the differential equation analytically, then we would have an equation of the form

$$y = F(x),$$

describing that curve; however, that is not how we would solve it on a computer.

Our next question is this: just how does the differential equation

$$\frac{dy}{dx} = f(x, y)$$

describe the solution $y = F(x)$? Since dy/dx is the slope of the curve $y = F(x)$, the differential equation gives us the direction of the solution at every point. Given any point (x, y) *on the curve*, we can obtain the direction of the curve through that point.

Leaving mathematics for a moment, we could interpret the problem in terms of following a map in search of a hidden treasure. Having an analytic solution of the differential equation is analogous to having a map of the area. If we then want the treasure, we look at the map and go directly there. A numerical solution, on the other hand, is analogous to having a starting point (always some old tree in the mystery stories!) and a set of directions. Using the directions (and a compass) we could start from the starting point and follow the path to the desired point. But note that without the map (analytic solution) we cannot take any shortcuts, we must follow the path (curve) as closely as we can to get to the desired goal.

Returning to Figure 7-2, we could use the starting point (a, b) and the differential equation to find the value of y for any desired value of x. Thus, if we wish to know the value $F(\bar{x})$ we merely follow the curve until we get to the point whose x coordinate is $\bar{x}$, and then look at the resulting value of y. This is the essence of any computer solution of a differential equation.

Having decided how to tackle the problem of solving the differential equation, we can now proceed to Euler's method.

Since we are given that the solution passes through the initial value (a, b), we can draw a tangent through this point as shown in Figure 7-3. From the differential equation we can find the slope of this tangent as

$$\left. \frac{dy}{dx} \right|_{\substack{\text{at } x=a \\ y=b}} = f(a, b)$$

If the solution $F(x)$ is continuous and differentiable, then in a small neighborhood of the point (a, b) we can approximate the curve by its tangent; this is essentially the same as expanding the function $y = F(x)$ around the point (a, b) using the first two terms of a Taylor series.

Using this technique we can get to the point $(a + w, b + h)$ a short distance from (a, b). We again calculate the slope at the new point, and take another short step further. By continuously iterating this process we can go as far as

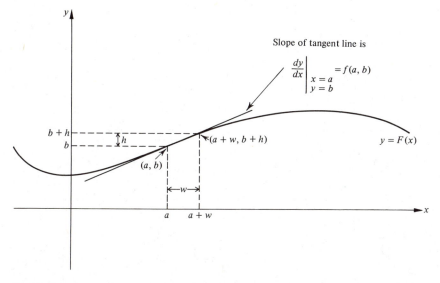

Figure 7-3.   Reconstructing the solution from its differential equation and the initial value.

we please. Unfortunately, as we shall see, roundoff error prevents us from getting even reasonable accuracy.

To establish the technique more accurately, let us consider the solution curve $y = F(x)$ shown in Figure 7-4, which is described by some differential equation of the form

$$\frac{dy}{dx} = f(x, y)$$

with the initial condition that $y = y_0$ when $x = x_0$. Let us suppose that we wish to find the value

$$\bar{y} = F(\bar{x}),$$

corresponding to the point $x = \bar{x}$ shown in Figure 7-4.

We start by dividing the interval from x_0 to $\bar{x}$ into n subintervals of width w,

$$w = \frac{\bar{x} - x_0}{n},$$

and then call the resulting boundaries between subintervals x_1, x_2, $\ldots$, x_{n-1}, x_n, so that $\bar{x}$ becomes x_n as shown. Note that this procedure bears a close resemblance to the first steps in performing a numerical integration.

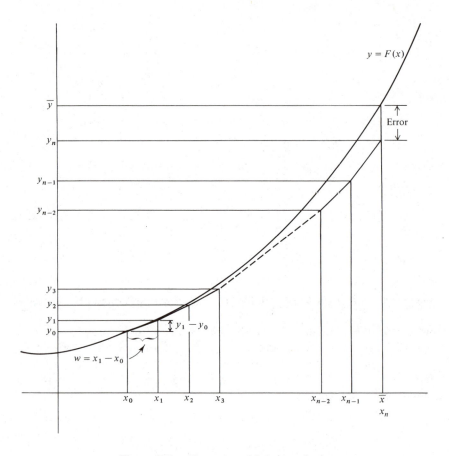

Figure 7-4.   Illustration of Euler's method.

Starting from the initial value (x_0, y_0) we draw a tangent to the curve with a slope

$$\frac{dy}{dx}\bigg|_{\substack{\text{at } x=x_0 \\ y=y_0}} = f(x_0, y_0).$$

If the y-coordinate of the tangent at the point $x = x_1$ is called y_1, then we can write

$$\text{Slope} = \frac{y_1 - y_0}{x_1 - x_0} = f(x_0, y_0),$$

and solve for y_1 as

$$y_1 - y_0 = f(x_0, y_0)(x_1 - x_0)$$
$$y_1 = y_0 + f(x_0, y_0)(x_1 - x_0). \qquad (7\text{-}1)$$

Since

$$y_0 = F(x_0),$$

and

$$f(x_0, y_0) = \left.\frac{dy}{dx}\right|_{\substack{\text{at } x = x_0 \\ y = y_0}} = F'(x_0),$$

we could also have written Eq. (7-1) as

$$F(x_1) \approx F(x_0) + F'(x_0)(x_1 - x_0), \qquad (7\text{-}2)$$

which recreates the first two terms of the Taylor series expansion.

Because this expression neglects the second- and higher-order derivatives of $F(x)$, y_1 as given here is not exactly equal to $F(x_1)$, resulting in a slight error.

Notice that the very first step along the curve results in an error. At the initial condition, x_0 and y_0 are specified exactly by the original problem, and so we can safely assume that

$$F(x_0) = y_0, \text{ exactly.}$$

But as soon as we make one step, we see that

$$F(x_1) \approx y_1, \text{ approximately.}$$

In order to minimize this error we might try to make $(x_1 - x_0)$ very small so that the tangent approximates the actual curve as closely as possible (and the remaining terms of the Taylor series approach zero). But this means that we will have to take many more steps to reach $\bar{x}$, thus increasing the total roundoff error. And so, when we finally finish, we may find a big difference between the calculated y_n and the correct $\bar{y} = F(\bar{x})$ as shown in Figure 7-4.

Thus in general we have an iteration of the form

$$y_1 = y_0 + f(x_0, y_0)w$$
$$y_2 = y_1 + f(x_1, y_1)w$$
$$y_3 = y_2 + f(x_2, y_2)w$$
$$\vdots$$
$$y_{i+1} = y_i + f(x_i, y_i)w \tag{7-3}$$
$$\vdots$$
$$y_n = y_{n-1} + f(x_{n-1}, y_{n-1})w$$

where $w = x_1 - x_0 = x_2 - x_1 = \cdots$ is the width of each interval, and is presumably very small.

As shown in Figure 7-4, we are essentially trying to approximate the curve $y = F(x)$ by a series of straight line segments. This leads to error unless we take very small intervals; actually, however, we cannot do this because of roundoff errors. In fact, there are three basic sources of error:

1. Since the approximation to a curve by a straight line cannot be exact, we have a basic error of the method caused by taking w too large. In similar fashion to previous chapters, we call this the truncation error. This error can be decreased as much as we like (at least theoretically) by merely taking w very small, but at the expense of time and roundoff error.
2. Increasing the number of subintervals with a very narrow width w involves many more calculations, and leads to some of the basic errors discussed in Chapter 1; the resulting roundoff error becomes large even with double-precision calculations.
3. As we can see from Figure 7-4, the calculation of each y_i is based on knowledge of the previous y_{i-1}. For example, in the next-to-last subinterval shown, we have

$$y_{n-1} = y_{n-2} + f(x_{n-2}, y_{n-2})w.$$

Quite obviously, any error in the y_{n-2} term will carry over into y_{n-1}, along with any new errors generated by roundoff in the present calculation. Another, not so obvious but potentially more dangerous error comes from the fact that an incorrect value of y_{n-2} is used in calculating $f(x_{n-2}, y_{n-2})$. Since this term indicates the direction of the curve, an error here causes us to step in the wrong direction, with the result that we may drift even further from the true curve.

This type of error (called *propagation* error) may cause slight errors made early in the process to be magnified later in the process. If $f(x, y)$ is strictly a function of x, then this effect should not be too serious. But if $f(x, y)$ is strongly dependent on y then every slight error in the value of y_{i-1} will strongly affect y_i.

When early errors in the process carry through but cause no further errors later, we say that the method is *stable*. If, on the other hand, early errors not only are carried over themselves, but cause other (often much larger) errors later, then we say that the method is *unstable*.

7.5 Example 7-1. Euler's Method

As we have already seen, if $y = k\, e^x$, then

$$\frac{dy}{dx} = k\, e^x = y,$$

and therefore the differential equation

$$\frac{dy}{dx} = y$$

is satisfied.

If, in addition, we are given the initial condition that for a particular $x = x_0$, $y = y_0$, then we can solve for the value of the constant k. For example, suppose we are given that $y = 1$ if $x = 0$. We then proceed as follows:

$$y = k\, e^x,$$

substituting $x = 0$, $y = 1$ into the equation, we have

$$1 = k\, e^0$$
$$= k \cdot 1$$
$$k = 1,$$

so that the specific solution is

$$y = F(x) = e^x.$$

By using Euler's method on the original differential equation, we could find e^x for any x. Suppose we wish to find the value of $e = e^1$. We then let $\bar{x} = 1$, and attempt to find

$$\bar{y} = F(\bar{x}) = e^{\bar{x}} = e^1 = e.$$

Since

$$\frac{dy}{dx} = y = f(x, y),$$

from the differential equation, we express the Euler iteration as

$$y_{i+1} = y_i + f(x_i, y_i)w \qquad (7\text{-}3)$$
$$= y_i + y_i w$$
$$= y_i(1 + w).$$

The following program uses this iteration by starting at the initial point (0, 1) and dividing the interval from $x = 0$ to $x = 1$ into $n = 1, 2, 4, \ldots,$ 16384 subintervals in turn.

```
      WRITE (3, 10)
   10 FORMAT ('    DIVISIONS        E          ERROR')
      DO 100 K = 1, 15
      N = 2 ** (K - 1)
      Y = 1.
      W = 1. / N
      DO 50 I = 1, N
   50 Y = Y * (1. + W)
      ERROR = Y - 2.718281
  100 WRITE (3, 110) N, Y, ERROR
  110 FORMAT (I8, 2F15.6)
      CALL EXIT
      END
```

Running the program, using standard precision arithmetic (equivalent to about 7 decimal digits), we obtain the following results:

DIVISIONS	E	ERROR
1	2.000000	-0.718280
2	2.250000	-0.468280
4	2.441406	-0.276874
8	2.565784	-0.152496
16	2.637927	-0.080354
32	2.676983	-0.041297
64	2.697332	-0.020948
128	2.707707	-0.010573
256	2.712931	-0.005349
512	2.715500	-0.002780
1024	2.716705	-0.001575
2048	2.717119	-0.001162
4096	2.716937	-0.001343
8192	2.716088	-0.002192
16384	2.714149	-0.004131

The closest answer occurs with 2048 subintervals, and even then only provides three-digit accuracy. This is an excellent example of why Euler's method is not too popular.

7.6 Extended Euler's Method

We have already pointed out that Euler's method uses the first two terms of a Taylor series for its iteration. We have written.

$$F(x_1) \approx F(x_0) + F'(x_0)(x_1 - x_0). \qquad (7\text{-}2)$$

Since the value of y_1 is obtained from the right side of the above equation, we could write

$$F(x_1) \approx F(x_0) + F'(x_0)(x_1 - x_0) = y_1,$$

which again points out that y_1 is not exactly equal to $F(x_1)$; in fact, we could use the Taylor series remainder term R_2 (for two terms) to calculate the error bound.

We are now tempted to expand again to find $F(x)$ as follows:

$$F(x_2) \approx F(x_1) + F'(x_1)(x_2 - x_1),$$

which, by itself, is approximately correct. But if we suppose that the right side is equal to y_2, we are *wrong*. As it turns out, the right side is absolutely useless in calculating y_2, because it requires knowledge of $F(x_1)$ and $F'(x_1)$, which we have no way of determining. Only in the first iteration, to find y_1, do we actually use a Taylor series expansion of $F(x)$, since we are given correct values for $y_0 = F(x_0)$ as an initial condition. After that we are reduced to using equations of the form

$$y_{i+1} = y_i + f(x_i, y_i)(x_{i+1} - x_i),$$

which only bear a superficial resemblance to the original Taylor series expansion.

Nevertheless, recognizing the resemblance, we can use our knowledge of Taylor series to improve the accuracy.

If we expand the Taylor series for $F(x_1)$ in three terms instead of two, we have

$$F(x_1) \approx F(x_0) + F'(x_0)(x_1 - x_0) + F''(x_0)\frac{(x_1 - x_0)^2}{2!}.$$

For the first iteration, this becomes[1]

$$y_1 = y_0 + f(x_0, y_0)(x_1 - x_0) + \frac{d}{dx}[f(x_0, y_0)]\frac{(x_1 - x_0)^2}{2},$$

---

[1] The notation $(d/dx)[f(x_0, y_0)]$ should be interpreted as $(d/dx)[f(x, y)]$ evaluated at $x = x_0$, and $y = y_0$.

since

$$F''(x) = \frac{d}{dx}F'(x) = \frac{d}{dx}[f(x, y)],$$

or, since w is the width of each interval,

$$y_1 = y_0 + f(x_0, y_0)w + \frac{d}{dx}[f(x_0, y_0)]\frac{w^2}{2}.$$

We now suspect that using the same iteration formula throughout will improve our accuracy, and so we adopt the iteration

$$y_{i+1} = y_i + f(x_i, y_i)w + \frac{d}{dx}[f(x_i, y_i)]\frac{w^2}{2}.$$

The usefulness of this method now depends on just how easily we can perform the required differentiation. If $f(x, y)$ is a function of x only, then differentiation with respect to x is relatively easy, and quite practical.

If, on the other hand, $f(x, y)$ is a function of y as well as x (which is the general case), then we have to use total derivatives. Assuming that $f(x, y)$ has derivatives of all orders and that they are differentiable, we can take the total derivative if we notice that y is a function of x:

$$\frac{d}{dx}[f(x, y)] = \frac{\partial f(x, y)}{\partial x} + \frac{\partial f(x, y)}{\partial y}\frac{dy}{dx},$$

where we are fortunate to know the value of dy/dx from the original differential equation.

Quite obviously this procedure is lengthy, but at least generally possible. Nevertheless, we learn that extension of the idea to use of four, five, or even more terms of the Taylor series is likely to involve even more complicated differentiation of $f(x, y)$, and thus be even less practical. The second derivative of $f(x, y)$ (needed if we wanted to use four terms of the series) is

$$\frac{d^2}{dx^2}[f(x, y)] = \frac{\partial^2 f(x, y)}{\partial x^2} + 2\frac{dy}{dx}\frac{\partial^2 f(x, y)}{\partial x\, \partial y}$$

$$+ \left(\frac{dy}{dx}\right)^2\frac{\partial^2 f(x, y)}{\partial y^2} + \frac{\partial f(x, y)}{\partial x}\frac{\partial f(x, y)}{\partial y}$$

$$+ \left(\frac{\partial f(x, y)}{\partial y}\right)^2\frac{dy}{dx},$$

and the higher derivatives are even more lengthy.

Since the use of additional terms from the Taylor series is so difficult, much work has gone into developing other methods equally good, but easier. In fact, one yardstick for evaluating them is to compare the number of derivatives needed in a Taylor series for the same accuracy. The standard Euler method, for example, is equivalent to taking terms through the first derivative, and is therefore called a *first-order* method.

7.7 Example 7-2. Extended Form of Euler's Method

If the required differentiation of $f(x, y)$ with respect to x is easy, then the use of the extended Euler's method can be quite simple. Thus in the differential equation of Example 7-1,

$$\frac{dy}{dx} = f(x, y) = y,$$

we have

$$\frac{d}{dx} f(x, y) = \frac{d}{dx}(y) = \frac{dy}{dx} = y.$$

The iteration therefore becomes quite simply

$$y_{i+1} = y_i + y_i w + y_i \frac{w^2}{2}$$

$$= y_i \left(1 + w + \frac{w^2}{2} \right).$$

This requires only one small change in the program of Example 7-1, the change of statement 50 to

```
50 Y = Y * (1. + W + W**2/2.)
```

which then gives the following results:

DIVISIONS	E	ERROR
1	2.500000	-0.218280
2	2.640625	-0.077655
4	2.694856	-0.023425
8	2.711840	-0.006440
16	2.716590	-0.001690
32	2.717843	-0.000437
64	2.718158	-0.000123
128	2.718225	-0.000056
256	2.718212	-0.000069
512	2.718146	-0.000134
1024	2.718022	-0.000258
2048	2.717760	-0.000521
4096	2.717249	-0.001031
8192	2.716228	-0.002052
16384	2.714149	-0.004131

By taking the second derivative of $F(x)$, the extended method gives us substantially better accuracy than the basic Euler method. With 128 subintervals, we have five-digit accuracy, whereas the standard Euler method gave only three-digit accuracy at 2048 intervals. The reason for the increased accuracy with fewer subintervals is the same as in integration; the truncation error decreases very quickly, and permits a lower error to be reached before roundoff error becomes severe. Note that the error at 16,384 subintervals is the same in both examples; it is mostly roundoff error, and has little to do with truncation error.

Hence extending the Euler method by taking one more term (or perhaps even more) is a convenient way of increasing the accuracy *if* we can easily differentiate the original differential equation to give $d/dx[f(x, y)]$. Unfortunately, this can be very difficult.

7.8 Modified Euler's Method

As we saw, the extended form of Euler's method gave us substantially better accuracy than the standard Euler method, but it cannot always be used. We would therefore like to find a simpler way of accomplishing the same result.

The standard Euler method provides inaccurate results because it tries to use the derivative found at one end of an interval as though it applied throughout the interval. To get reasonable accuracy we must use a very small interval so that the derivative is fairly constant over the interval, but this introduces a larger roundoff error.

The modified Euler's method tries to avoid this problem by using the *average* value of the derivative, instead of the derivative at one end.

The modified Euler method consists of two steps:

1. As before, we start with a point (x_i, y_i), and use the standard Euler method to compute a new $\overline{y_{i+1}}$ corresponding to the point $x = x_{i+1}$. In this part, we obtain the same $\overline{y_{i+1}}$ as we would obtain with the standard Euler method, but we are over-scoring the symbol since it is not the same as the final value we will get. This part of the process is called the *predictor* step, since it attempts to predict the next point on the curve from previous data.

2. The second step is called the *corrector* since it tries to correct the prediction. Having obtained a new point $(x_{i+1}, \overline{y_{i+1}})$ we now compute the derivative $f(x_{i+1}, \overline{y_{i+1}})$ at this point from the original differential equation, and take the average of this derivative with the derivative at the original point (x_i, y_i). We now *throw away the temporary point at* $(x_{i+1}, \overline{y_{i+1}})$ and return to (x_i, y_i). Using the *average* derivative, we now calculate

a new value y_{i+1} which should be more accurate than the predicted value $\overline{y_{i+1}}$.

This method is the simplest of an entire series of methods called *predictor-corrector* methods. Before we proceed, let us try a simple numerical example as an illustration, since this will demonstrate a possible pitfall.

In Example 7-1 we used the differential equation

$$\frac{dy}{dx} = y = f(x, y),$$

with the initial condition that $y = 1$ when $x = 0$ to find the value of e. If we were to use only one subinterval with the standard Euler method to find y at $x = 1$, we would use the iteration formula.

$$\overline{y_1} = y_0 + f(x_0, y_0)(x_1 - x_0)$$
$$= y_0 + y_0 \cdot 1 = 2y_0 = 2.$$

Using the standard, unmodified Euler method, we would now stop and claim that $\overline{y_1} = 2$ is the answer with, of course, quite an error. Instead, we now proceed with the corrector step by finding the new derivative

$$\frac{dy}{dx}\bigg|_{\substack{\text{at } x = x_1 \\ y = \overline{y_1}}} = f(x_1, \overline{y_1}) = f(1, 2) = \overline{y_1} = 2.$$

Averaging with the derivative $dy/dx = 1$ at the starting point $(0, 1)$, we have an average derivative of 1.5 in the interval. We now return to the point (x_0, y_0) and recompute a new y_1 from Eq. (7-1)

$$y_1 = y_0 + 1.5(x_1 - x_0) = 1 + 1.5 = 2.5,$$

where the average derivative is used instead of the $f(x_0, y_0)$ used previously. The new value of y_1 is substantially better than the originally predicted value of $\overline{y_1} = 2$.

We now encounter the promised pitfall. Noting that one correction step has substantially bettered our result, we immediately jump to the (hasty) conclusion that if one corrector step is good, two, three, or five are better. If we proceed along this path, we note that the derivative at the corrected value of $y_1 = 2.5$ is also 2.5, so that the average derivative over the interval is 1.75, and the second corrected value for y_1 is now

$$y_1 = y_0 + 1.75(x_1 - x_0) = 2.75,$$

which happens to overshoot the correct value of 2.718 .... If the step size w is too large, further iteration of the corrector step sometimes increases the over-all error, and may even cause divergence.

Assuming that the corrector step is done only once, let us formalize the modified Euler method. First, using the predictor step we obtain

$$\overline{y_{i+1}} = y_i + f(x_i, y_i)w.$$

Having obtained $\overline{y_{i+1}}$, we now calculate $f(x_{i+1}, \overline{y_{i+1}})$ which is the derivative at the new point $(x_{i+1}, \overline{y_{i+1}})$, and average with the previous derivative $f(x_i, y_i)$ to find the average derivative

$$\tfrac{1}{2}[f(x_i, y_i) + f(x_{i+1}, \overline{y_{i+1}})].$$

Now we substitute this average value into the original iteration equation instead of $f(x_i, y_i)$ to get

$$y_{i+1} = y_i + \frac{w}{2}[f(x_i, y_i) + f(x_{i+1}, \overline{y_{i+1}})].$$

7.9 Example 7-3. Use of the Modified Euler Method

The following program shows how we would use the modified Euler method to solve the differential equation

$$\frac{dy}{dx} = y$$

of Example 7-1, with the initial condition that $y = 1$ at $x = 0$. It differs from the program in Example 7-1 only in that the calculation of each new y_{i+1} consists first of a prediction to get the approximate value $\overline{y_{i+1}}$ (which is called YNEW in the program), then an averaging of the derivatives, and finally a correction to find a more accurate y_{i+1}. As before, the program is repeated for $n = 1, 2, 4, \ldots, 16384$ intervals.

```
        WRITE (3, 10)
     10 FORMAT ('   DIVISIONS      E          ERROR')
        DO 100 K = 1, 15
        N = 2 ** (K - 1)
        Y = 1.
        W = 1. / N
        DO 50 I = 1, N
C       PREDICTION
        YNEW = Y + Y * W
```

```
C       AVERAGE DERIVATIVE
        AVERG = (Y + YNEW) / 2.
C       CORRECTION
   50 Y = Y + AVERG * W
      ERROR = Y - 2.718281
  100 WRITE (3, 110) N, Y, ERROR
  110 FORMAT (I8, 2F15.6)
      CALL EXIT
      END
```

The modified Euler method gives us the following results:

DIVISIONS	E	ERROR
1	2.500000	-0.218280
2	2.640625	-0.077655
4	2.694856	-0.023425
8	2.711840	-0.006440
16	2.716590	-0.001690
32	2.717843	-0.000437
64	2.718158	-0.000123
128	2.718225	-0.000056
256	2.718212	-0.000068
512	2.718148	-0.000133
1024	2.718027	-0.000253
2048	2.717766	-0.000514
4096	2.717260	-0.001020
8192	2.716253	-0.002027
16384	2.714230	-0.004051

7.10 Integration and the Euler Method

We have already shown that using the modified predictor-corrector form of the Euler method involves using an iteration formula of the form

$$y_{i+1} = y_i + \frac{w}{2}[f(x_i, y_i) + f(x_{i+1}, \overline{y_{i+1}})].$$

By remembering trapezoidal integration from Chapter 6, we should recognize the term

$$\frac{w}{2}[f(x_i, y_i) + f(x_{i+1}, \overline{y_{i+1}})],$$

as being similar to trapezoidal integration.

Assuming that the analytic solution of our differential equation is

$$y = F(x),$$

we can write

$$\frac{dy}{dx} = F'(x) = f(x, y).$$

The only requirement is that, in order to find the derivative from $f(x, y)$, we must know not only the value of x where we want that derivative, but also the value of y at that point; this is generally not easy, but in no way prevents us from writing the above equation.

But now we can take

$$F'(x) = f(x, y)$$

and integrate both sides with respect to x:

$$\int F'(x)\, dx = F(x) = \int f(x, y)\, dx.$$

Knowing that $F(x)$ is the indefinite integral of $f(x, y)$ we can now integrate $f(x, y)$ between limits of $x = x_i$ and $x = x_{i+1}$ to get

$$\int_{x_i}^{x_{i+1}} f(x, y)\, dx = F(x_{i+1}) - F(x_i)$$

$$\approx y_{i+1} - y_i, \qquad (7\text{-}4)$$

where we note that, due to previous errors, y_i and y_{i+1} are not exactly equal to $F(x_i)$ and $F(x_{i+1})$.

On the other hand, we might try the same integration by using a trapezoidal approximation between the point (x_i, y_i), and the point $(x_{i+1}, \overline{y_{i+1}})$ obtained during the predictor step,

$$\int_{x_i}^{x_{i+1}} f(x, y)\, dx \approx \frac{w}{2}[f(x_i, y_i) + f(x_{i+1}, \overline{y_{i+1}})], \qquad (7\text{-}5)$$

where w is the width of the trapezoid,

$$w = x_{i+1} - x_i.$$

Equating the two integrals in Eqs. (7-4) and (7-5), we have

$$y_{i+1} - y_i \approx \frac{w}{2}[f(x_i, y_i) + f(x_{i+1}, \overline{y_{i+1}})]$$

$$y_{i+1} \approx y_i + \frac{w}{2}[f(x_i, y_i) + f(x_{i+1}, \overline{y_{i+1}})],$$

which brings us right back to the modified Euler method.

Having identified the modified Euler method with trapezoidal integration, we can now proceed to an extension which is based on Simpson's rule, and is called Milne's method.

7.11 Milne's Method

In the modified Euler method we used a predictor-corrector method which used the predictor equation

$$\overline{y_{i+1}} = y_i + f(x_i, y_i)w$$

to give a preliminary value for $\overline{y_{i+1}}$, and the corrector equation

$$y_{i+1} = y_i + \frac{w}{2}[f(x_i, y_i) + f(x_{i+1}, \overline{y_{i+1}})]$$

to give a final value for y_{i+1}. We also noted that the bracket term in the corrector represented a trapezoidal integration. Let us now derive a corrector based on Simpson's method. Noting that if $y = F(x)$ is the solution, then

$$\frac{dy}{dx} = \frac{d}{dx}F(x) = f(x, y),$$

so that $f(x, y)$ is the derivative of $F(x)$, then $F(x)$ is the integral of $f(x, y)$ and we can write

$$\int_{x_{i-1}}^{x_{i+1}} F'(x)\,dx = \int_{x_{i-1}}^{x_{i+1}} f(x, y)\,dx = F(x_{i+1}) - F(x_{i-1})$$

$$\approx \frac{w}{3}[f(x_{i-1}, y_{i-1}) + 4f(x_i, y_i) + f(x_{i+1}, y_{i+1})].$$

If we now assume that

$$y_{i+1} \approx F(x_{i+1})$$
$$y_{i-1} \approx F(x_{i-1}),$$

we can write the corrector formula as

$$y_{i+1} \approx y_{i-1} + \frac{w}{3}[f(x_{i-1}, y_{i-1}) + 4f(x_i, y_i) + f(x_{i+1}, \overline{y_{i+1}})],$$

where we again have a $\overline{y_{i+1}}$ on the right side, which must somehow be obtained from a predictor.

Since we still need a predicted value of $\overline{y_{i+1}}$, we might use one step of Euler's method, but Milne also derived the following predictor equation to give the predictor-corrector pair known as Milne's method:

Predictor

$$\overline{y_{i+1}} = y_{i-3} + \frac{4w}{3}[2f(x_i, y_i) - f(x_{i-1}, y_{i-1}) + 2f(x_{i-2}, y_{i-2})]$$

Corrector

$$y_{i+1} = y_{i-1} + \frac{w}{3}[f(x_{i-1}, y_{i-1}) + 4f(x_i, y_i) + f(x_{i+1}, y_{i+1})].$$

One interesting aspect of Milne's method is that it requires knowledge of four preceding values of y_i. Unlike Euler's method where the calculation of y_{i+1} only depends on y_i, here we need y_{i-1}, y_{i-2}, and y_{i-3} as well. In other words, Milne's method is not self-starting; to start using it, we must first use some other method to find y_1, y_2, and y_3 (where we are given y_0). Once we have done this, we can then use Milne's method to find y_4 and then continue. For this reason, Milne's method is called a *multi-step* method, as opposed to Euler's method which is called *one-step* or *single-step*.

Because of the need to provide a special starting method, and even more important, because it is inherently unstable, Milne's method (and specifically, the corrector) is not too popular. Still, as we shall see in the following example, when it works it gives rather good results.

7.12 Example 7-4. Use of Milne's Method

As in the previous examples of this chapter, we again seek to solve the differential equation

$$\frac{dy}{dx} = y,$$

subject to the initial condition that $F(0) = 1$, to find $F(1)$. Since the solution of the equation is $y = e^x$, the answer should be equal to $e = 2.718 \ldots$.

Since Milne's method needs four previous points, and y_0 is given, the program first uses the modified Euler method[2] to compute y_1, y_2, and y_3. After these values are determined, Milne's method takes over and computes the

---

[2] A more accurate, and more popular, method for starting Milne's method and calculating y_1, y_2, and y_3 is the Runge-Kutta method described later in this chapter.

remaining points. The program is repeated for 4, 8, 16, ... up to 16,384 divisions to give us an idea of the dependence of the results on the number of divisions and the interval width.

```
C     MILNE'S METHOD
      WRITE (3, 10)
   10 FORMAT ('   DIVISIONS        E               ERROR')
      DO 100 K = 3, 15
      N = 2 ** (K - 1)
      Y0 = 1.
      W = 1. / N
C                ------
C     NOTATION - Y(I+1) IS CALLED YTEMP
C                Y(I+1)            YNEXT
C                Y(I)             YNOW
C                Y(I-1)           YMIN1
C                Y(I-2)           YMIN2
C                Y(I-3)           YMIN3
C
C     GET Y(1), Y(2), AND Y(3) BY MODIFIED EULER METHOD
      Y1 = Y0 + W/2. * (Y0 + (Y0 + Y0 * W))
      Y2 = Y1 + W/2. * (Y1 + (Y1 + Y1 * W))
      Y3 = Y2 + W/2. * (Y2 + (Y2 + Y2 * W))
C     NEXT WE SET UP THE PREVIOUS VALUES NEEDED TO START MILNE'S METHOD
      YMIN3 = Y0
      YMIN2 = Y1
      YMIN1 = Y2
      YNOW  = Y3
C     FINALLY WE CAN CONTINUE WITH MILNE'S METHOD
      LASTI = N - 1
      DO 50 I = 3, LASTI
C     PREDICTOR
      YTEMP = YMIN3 + 4. * W/3. * (2. * YNOW - YMIN1 + 2. * YMIN2)
C     CORRECTOR
      YNEXT = YMIN1 + W/3. * (YMIN1 + 4. * YNOW + YTEMP)
C     NOW PREPARE FOR THE NEXT ITERATION
      YMIN3 = YMIN2
      YMIN2 = YMIN1
      YMIN1 = YNOW
   50 YNOW  = YNEXT
      ERROR = YNEXT - 2.718281
  100 WRITE (3, 110) N, YNOW, ERROR
  110 FORMAT (I8, 2F15.6)
      CALL EXIT
      END
```

When we run the preceding program using Milne's method, we obtain the following results:

DIVISIONS	E	ERROR
4	2.705266	−0.013014
8	2.716431	−0.001850
16	2.718035	−0.000246
32	2.718247	−0.000033
64	2.718270	−0.000010
128	2.718264	−0.000016
256	2.718250	−0.000030
512	2.718217	−0.000063
1024	2.718150	−0.000131
2048	2.718026	−0.000255
4096	2.717772	−0.000508
8192	2.717264	−0.001016
16384	2.716254	−0.002027

Comparing these results with the modified Euler method of Example 7-3, we see that the minimum error (in this particular case) occurs with only 64 divisions instead of 128, and the accuracy is five times better—an error of only -0.000010 instead of -0.000056. But, as expected, there is a price to be paid for this in terms of increased complexity and increased time needed for solution.

7.13 Iteration Errors and Hamming's Correction

So far in our discussion of various methods for solving ordinary differential equations, we have very carefully avoided any quantitative discussion of error, and for a good reason.

As pointed out earlier in this chapter, there are three sources of errors in the numerical solution of differential equations:

1. Truncation error, which could be called the "error of the method". This error is relatively easy to analyze at each step of the iteration, although the overall truncation error of the entire solution is a bit less clear.
2. Roundoff error, which is caused by having a very large number of calculations to do, either because the width w is very small, or because we are using a very complicated set of equations. The analysis of roundoff error could also be done, using techniques similar to those of Chapter 1, if we absolutely had to.
3. Propagation error is introduced because the calculation of $f(x, y)$ at each iteration depends on the results of previous calculations. Thus previous errors not only carry over into the final answer, but also introduce new errors of their own.

Thus the over-all error of a solution depends on three sources of error of which one, propagation error, is extremely difficult to analyze or even estimate. We cannot even hope to estimate total error without first somehow tackling propagation error, and we shall not even attempt to do so.

On the other hand, we can intuitively try to decrease propagation error by noting that it is caused by other errors. If we could reduce these other errors, we would reduce propagation errors as well. Since there is little we can do about roundoff error except by using double precision or a different computer which uses longer words and hence has greater accuracy, we can only try to cut truncation error. One way is to use a more complex method of prediction-correction or some other scheme; another is to try to guess what our errors are and correct for them. This is the method suggested by Hamming,[3] which we shall apply to Milne's method.

[3] R. W. Hamming, "Stable Predictor-Corrector Methods for Ordinary Differential Equations," *Journal of the ACM*, **6**, pp. 37–47 (1959).

As pointed out earlier, the corrector of Milne's method uses a Simpson integration. As we have already found in Chapter 6, the error in one application of Simpson's rule is about

$$E_i = A_i - I_i \approx \tfrac{1}{90} f^{IV}(x_i) w^5,$$

where, to refresh our memories,

I_i is the mathematically exact value for the integral in the ith segment of the integration;

A_i is the integral as found from Simpson's rule;

E_i is the resulting error;

$f^{IV}(x_i)$ is the fourth derivative of the function $f(x)$; evaluated at $x = x_i$;

w is the width of the integration interval.

Rearranging the error equation, we have

$$I_i \approx A_i - \tfrac{1}{90} f^{IV}(x_i) w^5.$$

Now, in going from Chapter 6 to Chapter 7 we have kept our notation the same. In Chapter 6 we integrated a function $f(x)$ to get $F(x)$, so that $f(x)$ was the derivative of $F(x)$. In this chapter also we have $f(x, y)$ as the derivative of $F(x)$, and the only difference is that the $f(x)$ of Chapter 6 now becomes a function of both x and y, $f(x, y)$. Otherwise, we may apply the Simpson's rule error equation to the Milne's method corrector to get

$$\text{True } F(x_{i+1}) \approx \text{Calculated } y_{i+1} - \tfrac{1}{90} f^{IV}(x_i, y_i) w^5,$$

where now $f^{IV}(x_i, y_i) = (\partial^4/\partial x^4) f(x, y)$ evaluated at (x_i, y_i).

In a process similar to the derivation of the Simpson's rule error term in Chapter 6, we could next use our knowledge of Taylor series to determine the error in the Milne predictor equation. If we did so, we would obtain

$$\text{True } F(x_{i+1}) \approx \text{Calculated } \overline{y_{i+1}} + \tfrac{28}{90} f^{IV}(x_i, y_i) w^5.$$

For any particular iteration, let $K = f^{IV}(x_i, y_i) w^5$, so that the two equations become

$$F(x_{i+1}) \approx y_{i+1} - \tfrac{1}{90} K \qquad (7\text{-}6)$$

$$F(x_{i+1}) \approx \overline{y_{i+1}} + \tfrac{28}{90} K.$$

By equating the two, we can solve for K

$$\overline{y_{i+1}} + \tfrac{28}{90}K = y_{i+1} - \tfrac{1}{90}K$$

$$y_{i+1} - \overline{y_{i+1}} = \tfrac{28}{90}K + \tfrac{1}{90}K = \tfrac{29}{90}K$$

$$K = \tfrac{90}{29}[y_{i+1} - \overline{y_{i+1}}].$$

Inserting this value of K into the corrector equation, Eq. (7-6), we have

$$F(x_{i+1}) \approx y_{i+1} - \tfrac{1}{90}\tfrac{90}{29}[y_{i+1} - \overline{y_{i+1}}]$$
$$\approx y_{i+1} - \tfrac{1}{29}[y_{i+1} - \overline{y_{i+1}}].$$

This error correction due to Hamming therefore states that after we calculate y_{i+1} from the corrector equation, we should subtract an extra correction term equal to $\tfrac{1}{29}$ of the difference between the predictor and corrector result. The result should (hopefully) be closer to the correct answer, although there are still various approximations made which will stop us from getting an exact correction.

Lest the student think that such a correction term is a cure-all for his problems, we present here the results as applied to the $dy/dx = y$ example of this chapter. As can be seen, the gain in accuracy in this particular example is rather small; in fact, to make it apparent we had to perform the entire calculation in double-precision. As you can see also, the correction term has no effect on the answer with many divisions, since the difference between the predictor and corrector using a complex iteration equation such as Milne's is fairly small, and the correction term itself is therefore insignificant; the roundoff error greatly exceeds the truncation error term for which we are correcting.

Milne's method without Hamming's correction:

DIVISIONS	E	ERROR
4	2.705266	-0.130155E-01
8	2.716431	-0.185040E-02
16	2.718035	-0.245904E-03
32	2.718250	-0.317208E-04
64	2.718277	-0.405125E-05
128	2.718281	-0.568106E-06
256	2.718281	-0.199303E-06
512	2.718281	-0.249594E-06
1024	2.718281	-0.497326E-06
2048	2.718280	-0.989064E-06
4096	2.718279	-0.197067E-05
8192	2.718277	-0.395439E-05
16384	2.718274	-0.782869E-05

Milne's method with Hamming's correction:

DIVISIONS	E	ERROR
4	2.705400	-0.128809E-01
8	2.716457	-0.182441E-02
16	2.718040	-0.241434E-03
32	2.718250	-0.310633E-04
64	2.718277	-0.394694E-05
128	2.718281	-0.556930E-06
256	2.718281	-0.199303E-06
512	2.718281	-0.249594E-06
1024	2.718281	-0.497326E-06
2048	2.718280	-0.989064E-06
4096	2.718279	-0.197067E-05
8192	2.718277	-0.395439E-05
16384	2.718274	-0.782869E-05

7.14 Runge-Kutta Methods

As pointed out earlier, Euler's method is called a single-step method since it relies only on knowledge of (x_i, y_i) in computing (x_{i+1}, y_{i+1}); Milne's method, on the other hand, is called a multi-step method since it requires knowledge of several previous points. In a way, the Runge-Kutta methods are a combination of the two and combine the better points of each.

The advantage of the Runge-Kutta methods over Euler's method is vastly greater accuracy for the same (or comparable) amount of calculations; their advantage over Milne's method is partially their stability, and partially that they are self-starting and do not require some other method to provide the first few points. On the other hand, the analysis of errors in the Runge-Kutta methods is even less obvious than in the other methods. Furthermore, a good predictor-corrector can often produce better accuracy.

All of the Runge-Kutta methods are derived from a Taylor series in such a way that the benefits of expanding the function in a series are preserved without the actual need for finding many derivatives. Depending on the number of terms taken from the Taylor series, we can get various Runge-Kutta methods, which differ from each other in their order (we have defined the order of a method for solving an ordinary differential equation as the number of terms needed in a Taylor series for the same accuracy.) There are Runge-Kutta methods of order 2, 3, 4, 5, and so on, although the fourth-order (also called *quartic*) method seems to be most popular. Its derivation is somewhat long and difficult, and so we will confine ourselves to stating the results.

As before, the region from the initial value point x_0 to the point of the desired solution at $\bar{x}$ is divided into n subintervals. In each interval, we perform an iteration of the form

$$y_{i+1} = y_i + \tfrac{1}{6}(k_1 + 2k_2 + 2k_3 + k_4),$$

where

$$k_1 = f(x_i, y_i)w$$
$$k_2 = f(x_i + \tfrac{1}{2}w, y_i + \tfrac{1}{2}k_1)w$$
$$k_3 = f(x_i + \tfrac{1}{2}w, y_i + \tfrac{1}{2}k_2)w$$
$$k_4 = f(x_i + w, y_i + k_3)w.$$

In a way, this method is a multi-step method since it uses the points at $x = x_i$, at $x = x_i + \tfrac{1}{2}w$, and at $x = x_i + w$ in its calculations, yet it generates its own intermediate steps instead of relying on previous steps from previous calculations.

This fourth-order method is quite popular since it is quite accurate, stable, and relatively easy to program.

7.15 Example 7-5. Use of the Quartic Runge-Kutta Method

The following program again seeks to obtain the value of the exponential constant e from the differential equation

$$\frac{dy}{dx} = y,$$

subject to the initial condition that $F(0) = 1$, as in previous examples. As before, it is repeated for $1, 2, 4, \ldots, 16{,}384$ divisions to display the dependence of accuracy on the number of divisions.

```
C       RUNGE-KUTTA METHOD
        REAL K1, K2, K3, K4
        WRITE (3, 10)
   10 FORMAT ('   DIVISIONS          E           ERROR')
        DO 100 K = 1, 15
        N = 2 ** (K - 1)
        Y = 1.
        W = 1. / N
        DO 50 I = 1, N
        K1 = W * Y
        K2 = W * (Y + K1/2.)
        K3 = W * (Y + K2 / 2.)
        K4 = W * (Y + K3)
   50 Y = Y + (K1 + 2. * K2 + 2. * K3 + K4) / 6.
        ERROR = Y - 2.718281
  100 WRITE (3, 110) N, Y, ERROR
  110 FORMAT (I8, 2F15.6)
        CALL EXIT
        END
```

We obtain the following results from the program:

DIVISIONS	E	ERROR
1	2.708333	-0.009947
2	2.717346	-0.000934
4	2.718209	-0.000071
8	2.718276	-0.000005
16	2.718277	-0.000003
32	2.718273	-0.000008
64	2.718264	-0.000016
128	2.718250	-0.000030
256	2.718217	-0.000063
512	2.718148	-0.000132
1024	2.718027	-0.000253
2048	2.717766	-0.000514
4096	2.717260	-0.001020
8192	2.716253	-0.002027
16384	2.714230	-0.004051

We note that the results of the method are substantially better than those of previous methods. Only 16 divisions are needed to reach minimum error, which itself is much lower than that of previous methods. For a fair comparison with other methods, we should note that four evaluations of $f(x, y)$ are needed in each loop.

As an aside, the problem of just how many divisions to use is more acute here than in Milne's method, since in Milne's method (or any other predictor-corrector scheme) the quantity $(y_{i+1} - \overline{y_{i+1}})$, which is the difference between the predicted and corrected values, provides at least a clue to the relative size of the step error; this clue is not available to us in the Runge-Kutta method. In any case, the question remains of how many intervals we dare to use before the roundoff error becomes significant.

The hint we gave in Chapter 6 for integration applies here as well, however. In essence, we perform the same calculation with both standard and double-precision arithmetic. As long as the two results are the same, we can be assured that roundoff error is rather small. As soon as the two values become different, we know that standard precision roundoff error is becoming serious, and that a short time later the double-precision error will also increase.

7.16 Second- and Higher-Order Differential Equations

First-order differential equations such as those of the main part of this chapter are fairly easy to solve, but are not necessarily the most common. We often encounter higher-order differential equations of the form

$$a_n \frac{d^n y}{dx^n} + a_{n-1} \frac{d^{n-1} y}{dx^{n-1}} + \cdots + a_1 \frac{dy}{dx} = g(x, y),$$

which is of order n. By a proper transformation, we can solve even these equations by the methods of this chapter. First, we note that every such equation can be rewritten in the form

$$\frac{d^n y}{dx^n} = f(x, y, y', y'', \ldots, y^{(n-1)}),$$

where the primes indicate derivatives of y with respect to x.

We next perform the following change of variable:

$$
\begin{aligned}
y_1 &= y \\
y_2 &= y_1' & = y' \\
y_3 &= y_2' & = y'' \\
y_4 &= y_3' & = y''' \\
&\;\;\vdots \\
y_n &= y_{n-1}' = y^{(n-1)} \\
& \quad\;\; y_n' = y^{(n)}
\end{aligned}
$$

If the original equation is of order n, we now generate n first-order equations as follows:

$$
\begin{aligned}
y_1' &= y_2 \\
y_2' &= y_3 \\
y_3' &= y_4 \\
&\;\;\vdots \\
y_{n-1}' &= y_n \\
y_n' &= y^{(n)} \\
&= f(x, y, y', y'', \ldots, y^{(n-1)}) \\
&= f(x, y_1, y_2, y_3, \ldots, y_n).
\end{aligned}
$$

We thus have n equations, each of which is only first-order. As an example, if we had the differential equation

$$\frac{d^2 y}{dx^2} + \frac{dy}{dx} = x^2 + y^2,$$

we would transform it into the two equations

$$
\begin{aligned}
y_1' &= y_2 \\
y_2' &= x^2 + y_1^2 - y_2,
\end{aligned}
$$

both of which are first-order differential equations similar to those studied earlier in this chapter.

In general, then, an nth-order differential equation becomes changed into n simultaneous differential equations of the general form

$$y'_1 = f_1(x, y_1, y_2, y_3, \ldots, y_n)$$
$$y'_2 = f_2(x, y_1, y_2, y_3, \ldots, y_n)$$
$$y'_3 = f_3(x, y_1, y_2, y_3, \ldots, y_n)$$
$$\vdots$$
$$y'_n = f_n(x, y_1, y_2, y_3, \ldots, y_n).$$

We can now solve these by applying the Runge-Kutta method to each equation, and by iterating each equation in turn, as we did with nonlinear simultaneous equations in Chapter 4.

As an example, the Runge-Kutta method applied to two simultaneous equations of the form

$$y'_1 = f_1(x, y_1, y_2) \quad \text{or} \quad y' = f_1(x, y, z)$$
$$y'_2 = f_2(x, y_1, y_2) \quad \text{or} \quad z' = f_2(x, y, z),$$

(where we have written y instead of y_1 and z instead of y_2 in order to eliminate double subscripts in the following equations) would result in the iteration equations

$$y_{i+1} = y_i + \tfrac{1}{6}(k_1 + 2k_2 + 2k_3 + k_4)$$
$$z_{i+1} = z_i + \tfrac{1}{6}(c_1 + 2c_2 + 2c_3 + c_4),$$

which are always performed alternately, and the k's and c's are calculated from

$$k_1 = f_1(x_i, y_i, z_i)w,$$
$$c_1 = f_2(x_i, y_i, z_i)w,$$
$$k_2 = f_1(x_i + \tfrac{1}{2}w, y_i + \tfrac{1}{2}k_1, z_i + \tfrac{1}{2}c_1)w,$$
$$c_2 = f_2(x_i + \tfrac{1}{2}w, y_i + \tfrac{1}{2}k_1, z_i + \tfrac{1}{2}c_1)w,$$
$$k_3 = f_1(x_i + \tfrac{1}{2}w, y_i + \tfrac{1}{2}k_2, z_i + \tfrac{1}{2}c_2)w,$$
$$c_3 = f_2(x_i + \tfrac{1}{2}w, y_i + \tfrac{1}{2}k_2, z_i + \tfrac{1}{2}c_2)w,$$
$$k_4 = f_1(x_i + w, y_i + k_3, z_i + c_3)w,$$
$$c_4 = f_2(x_i + w, y_i + k_3, z_i + c_3)w,$$

which are calculated in that order. For higher-order systems of equations it is easy to write out the additional Runge-Kutta iteration equations needed.

7.17 Problems

1. As discussed in Chapter 1, adding a relatively large number to a relatively small number will result in a small error. In Examples 7-1, 7-2, and 7-3 this can result in a rather large error when we start with a (relatively) large value of y_0 and then add successively smaller numbers representing the changes in the y coordinate. Change these programs to add the coordinate changes separately, and add to y_0 (equal to 1 in these examples, but possibly larger in other problems) only when necessary for the next calculation.

2. Given a differential equation of the form

$$\frac{dy}{dx} = f(x, y)$$

the accuracy of a numerical solution depends greatly on whether $f(x, y)$ is strongly dependent on y. Consider the two equations

     a. $\dfrac{dy}{dx} = y;$ $F(0) = 1$

     b. $\dfrac{dy}{dx} = e^x;$ $F(0) = 1,$

which both have $F(x) = e^x$ as the solution. All five examples of this Chapter solve differential equation a; rewrite them to solve equation b instead, and compare results. Note: For a valid comparison, both equations should be solved on the same computer using the same precision arithmetic. Hence, you will have to run the sample programs on your own computer.

3. Plot the direction fields for the two equations in problem 2, using Figure 7-1 as an example, and compare.

4. Repeat problems 2 and 3 for the following two differential equations:

     a. $\dfrac{dy}{dx} = 2x;$ $F(1) = 1$

     b. $\dfrac{dy}{dx} = \dfrac{2y}{x};$ $F(1) = 1.$

Find $F(3)$. (Note: for both equations, $F(x) = x^2$.)

5. Make appropriate changes in the examples from this chapter to find e^{-1} instead of e^{+1}. (Note: do not use $e^{-1} = 1/e$.)

6. Solve each of the following differential equations by Euler's method, modified Euler's method, Milne's method, and the quartic Runge-Kutta method:

	Differential Equation	Initial Condition	Solve for
a.	$\dfrac{dy}{dx} = \cos x$	$F(0) = 0$	$F\left(\dfrac{\pi}{2}\right)$
b.	$\dfrac{dy}{dx} = \dfrac{1}{x}$	$F(1) = 0$	$F(2)$
c.	$\dfrac{dy}{dx} = \dfrac{y}{x \ln x}$	$F(1) = 0$	$F(2)$
d.	$\dfrac{dy}{dx} = 3x^2$	$F(-1) = -1$	$F(1)$
e.	$\dfrac{dy}{dx} = 2y + \cos x - 2 \sin x$	$F(0) = 1$	$F(1)$

7. Another popular multi-step method, called the Adams-Bashforth method, uses the iteration equation

$$y_{i+1} = y_i + \frac{w}{24}(55k_1 - 59k_2 + 37k_3 - 9k_4),$$

where

$$k_1 = f(x_i, y_i)$$
$$k_2 = f(x_{i-1}, y_{i-1})$$
$$k_3 = f(x_{i-2}, y_{i-2})$$
$$k_4 = f(x_{i-3}, y_{i-3}).$$

Using the quartic Runge-Kutta method for starting and the Adams-Bashforth method for continuing, write a program which will repeat the problem of Example 7-1, and compare the results with the other examples of this chapter.

8. A popular corrector equation is the Adams-Moulton formula

$$y_{i+1} = y_i + \frac{w}{24}(9k_1 + 19k_2 - 5k_3 + k_4),$$

where

$$k_1 = f(x_{i+1}, \overline{y_{i+1}})$$
$$k_2 = f(x_i, y_i)$$
$$k_3 = f(x_{i-1}, y_{i-1})$$
$$k_4 = f(x_{i-2}, y_{i-2}).$$

Using the quartic Runge-Kutta method for starting, the Adams-Bashforth method (see problem 7) as a predictor and the Adams-Moulton method as a corrector, write a program which will repeat the problem of Example 7-1, and compare the results with the other examples of this chapter, and with the results of problem 7.

8

Interpolation and Curve Fitting

We are often given a set of discrete values of a function, either in the form of a table of values or a set of measurements which actually represent a set of points along a continuous function. Given these discrete values, we wish to find intermediate values.

If the given values of the function are reasonably exact, such as might be given in a book of trigonometric tables, the process of finding intermediate values is called *interpolation*. If, on the other hand, the given values are not very accurate, such as if they come from actual measurements, then we need to perform a different process called *curve-fitting*, which tries to average out some of the errors.

8.1 Examples of Interpolation and Curve-Fitting

As an example, let us look at the problem of finding the sine of 6.5 degrees. Let us assume that we are given a table of sines, where the values are given for steps of one degree. If our task is to find the sine of 6.5 degrees, we have three choices:

1. Use a Taylor series to calculate the sine to as much accuracy as we desire.

2. Try to find a book of tables which lists the sine in smaller steps of perhaps 0.5 or 0.1 degree, and look up the exact value.
3. Using the sines for 6 and 7 degrees given in the available table, try to interpolate for the value at 6.5 degrees.

This example illustrates several common characteristics of interpolation problems. We usually know exactly what function we are considering, and probably even have an explicit equation which could be used to find the answer (such as the Taylor series for the sine). Yet, for some reason such as lack of time, we choose not to use the explicit equation, preferring instead to use table-lookup. But if the available table does not include all the desired values of the argument, or if the required table having all the desired information is too long to fit into computer memory, we may have to use interpolation along with a less detailed table. Thus we are starting with a set of tabulated points which are (presumably) correct to some desired accuracy.

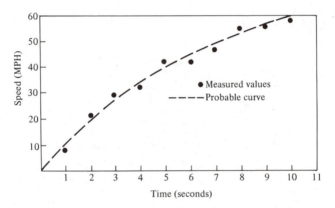

Figure 8-1.   Fitting a curve to a set of measurements.

In curve-fitting, on the other hand, we may not have such a set of exact points available. Consider the following example. While testing an automobile for its acceleration, the car is started from a halt, on a level road, and driven at maximum acceleration until its speed reaches 60 miles per hour. At the same time, speedometer readings are made at 1-second intervals. When the speed is plotted as a function of time as in Figure 8-1, we have a set of points.

We would of course expect these points to define a smooth curve. But due to errors in measurements and other factors, we see that the points are not so well arranged, so that some speed readings are too high and others are too low. Now, suppose that we wish to know the speed at 6.5 seconds. We could of course interpolate between the readings at 6 seconds and at 7 seconds, but since these are probably in error we would get a value which might not be nearly correct. What to do?

The solution to our problem is to try to fit a "probable curve" to the set of measured points. Since it is likely that the various measurements are inaccurate, this curve need not necessarily go through any of the points. On the other hand, since the errors in the measurements are probably not too large, the curve should at least pass near to every point, probably above some points, and below others.

Above all, the probable curve should be smooth. As we can see in Figure 8-2, we could devise a curve such that it would go through *all* the measured points, but it is unlikely that the automobile we are testing would really behave in such a jerky manner. The probable curve in Figure 8-1, on the other hand, seems much more plausible. Once we have such a curve, we can use it to determine the desired speed at 6.5 seconds, or make other measurements from it.

Thus the problem in curve-fitting is to find the most likely probable curve, one that provides the "best fit." There are different ways of defining a "best" fit, as we shall see later.

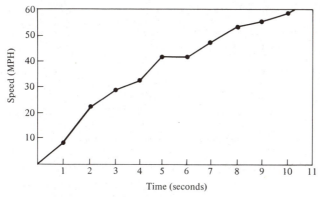

Figure 8-2. Is this the "best fit?"

8.2 Linear Interpolation

The simplest kind of interpolation is *linear* interpolation. Assuming some desired function $f(x)$, which is continuous and differentiable at all points, suppose we are given a table of values like this:

x	$f(x)$
x_0	$f(x_0)$
x_1	$f(x_1)$
x_2	$f(x_2)$
$\vdots$	$\vdots$
x_n	$f(x_n)$

Thus, for $n + 1$ different values of x, *not necessarily evenly spaced*, we are given the corresponding values of $f(x)$. We assume here that both the x_i and the corresponding $f(x_i)$ are given either exactly, or within some specified accuracy. Figure 8-3 shows the function $f(x)$ and the corresponding tabulated values, which are shown as heavy black points on the curve.

Now let us suppose we wish to know $f(\bar{x})$, where $\bar{x}$ is, for the sake of discussion, somewhere between x_3 and x_4, as shown in Figure 8-3.[1] To use linear interpolation, we draw a straight line between two tabulated points, one on each side[2] of the unknown point $\bar{x}$; in this case, we draw a straight line AD between the tabulated points at x_3 and x_4.

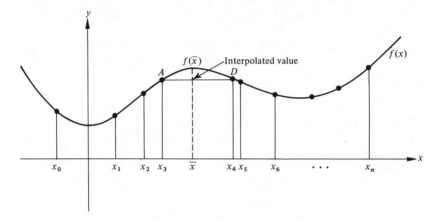

Figure 8-3.   Linear interpolation.

Having drawn this line, as in Figure 8-3, we now approximate the curve in the region between, in this case, x_3 and x_4 by the straight line, which is shown magnified in Figure 8-4. Using similar triangles, we form the proportion

$$\frac{BC}{AC} = \frac{DE}{AE},$$

which we can solve for BC:

$$BC = \frac{AC}{AE} DE$$

$$= \frac{\bar{x} - x_3}{x_4 - x_3}[f(x_4) - f(x_3)].$$

---

[1] The point $\bar{x}$ could be anywhere in the region between x_0 and x_n for interpolation.
[2] With some loss of accuracy, we could use two points both of which were to one side of the desired point; in this case, the process would be called *extrapolation*.

Our resulting interpolated value for $f(\bar{x})$ is then

$$p(\bar{x}) = f(\bar{x})_{int} = f(x_3) + \frac{\bar{x} - x_3}{x_4 - x_3}[f(x_4) - f(x_3)],$$

where $p(\bar{x})$ is the interpolating approximation to $f(\bar{x})$. In general, suppose we wish to find the value of $f(x)$ for some x located between x_i and x_{i+1} in the table of given values; then the interpolated value $p(x)$, which is only an approximation for $f(x)$, is given by

$$p(x) = f(x_i) + \frac{x - x_i}{x_{i+1} - x_i}[f(x_{i+1}) - f(x_i)]. \qquad (8\text{-}1)$$

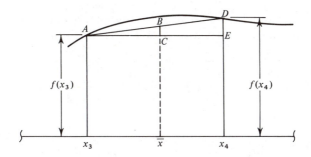

Figure 8-4.   Derivation of the linear interpolation formula.

As an example, suppose we try to find the sin of $6.5°$ from a book of tables which gives

$$\sin 6° = 0.10453$$
$$\sin 7° = 0.12187$$

We find

$$p(6.5) = f(6) + \frac{6.5 - 6}{7 - 6}[f(7) - f(6)]$$

$$= 0.10453 + \frac{0.5}{1}(0.01734)$$

$$= 0.10453 + 0.00867 = 0.11320,$$

which is correct within the five place accuracy limitations of the original table.

Quite obviously, if we want the best possible accuracy from such an interpolation, we should pick two points x_i and x_{i+1} which are close to the needed value of $\bar{x}$. As a simple example, if we had a table which had only sines for every 10 degrees, we would have to start from

$$\sin 0° = 0.0$$
$$\sin 10° = 0.17365,$$

and interpolate as follows:

$$p(6.5) = f(0) + \frac{6.5 - 0}{10 - 0}[f(10) - f(0)]$$

$$= 0.0 + 0.65(0.17365)$$

$$= 0.11287,$$

which has a substantial error. As another example, we might try to use the same equation and *extra*polate from, say, the starting values of

$$\sin 4° = 0.06976$$
$$\sin 5° = 0.08716,$$

as follows:

$$p(6.5) = f(4) + \frac{6.5 - 4}{5 - 4}[f(5) - f(4)]$$

$$= 0.06976 + 2.5(0.01740)$$

$$= 0.11326,$$

which again is in error.

Returning to the original interpolation formula,

$$p(x) = f(x_i) + \frac{x - x_i}{x_{i+1} - x_i}[f(x_{i+1}) - f(x_i)], \qquad (8\text{-}1)$$

we note that if $x = x_i$, then the second term on the right drops out, so that

$$p(x_i) = f(x_i),$$

and the result from the interpolating equation is exact. Similarly, if $x = x_{i+1}$, we have

$$p(x_{i+1}) = f(x_i) + f(x_{i+1}) - f(x_i)$$
$$= f(x_{i+1})$$

and again the result from the equation is exact. This is intuitively obvious from Figure 8-4; since the straight line approximation actually crosses the curve at these two points, we would expect the answer to be the same as the $f(x)$ given at those points. Because $p(x)$ exactly matches $f(x)$ at x_i and x_{i+1}, we say that $p(x)$ *interpolates exactly* at x_i and x_{i+1}.

Let us now rearrange the right-hand side of the interpolation formula into the form

$$p(x) = \left\{ \frac{f(x_{i+1}) - f(x_i)}{x_{i+1} - x_i} \right\} x + \left\{ f(x_i) - \frac{f(x_{i+1}) - f(x_i)}{x_{i+1} - x_i} x_i \right\}.$$

Once we choose the two tabular points x_i and x_{i+1}, the two quantities in brackets above are constants, which we could call a_1 and a_0; in this case, we can express $p(x)$ as

$$p(x) = a_1 x + a_0.$$

This form of the interpolating equation makes it obvious why we use the term *interpolating polynomial.* Indeed, the equation for linear interpolation is just a first-degree polynomial in x. For this reason, we call linear interpolation a first-degree interpolation.

8.3 The Interpolating Polynomial

In the previous section we discussed linear interpolation as being the process where we approximate a small portion of a curve with a straight line which passes through two known points of the curve; for best accuracy, we choose the two points as close to the desired point as possible, as shown in Figure 8-3.

As a better approximation, we might choose to approximate the function $f(x)$ with some higher-order curve. Just as two points on the curve define a unique straight line passing through them, so three points define a unique parabola, four points define a unique cubic curve, and so on. Choosing, say, m points out of the given $n + 1$ points, we could pass an $(m - 1)$ degree polynomial through these points. (The best we could do with $n + 1$ points is an nth degree approximation.) As in the linear case, an $(m - 1)$ degree interpolation would, after some simplifying, give us an $(m - 1)$ degree interpolating polynomial

$$p_{m-1}(x) = a_{m-1}x^{m-1} + a_{m-2}x^{m-2} + \cdots + a_2x^2 + a_1x + a_0. \quad (8\text{-}2)$$

As an example, suppose we were given a sine table with ten entries, giving sines from $0°$ to $90°$ in steps of 10 degrees. The best interpolation we could perform would be a 9th degree interpolating polynomial

$$p_9(x) = a_9 x^9 + a_8 x^8 + \cdots + a_2 x^2 + a_1 x + a_0.$$

Unlike the linear case, the coefficients in the higher-degree interpolating polynomial are not at all easy to find. To find them, we use the method of undetermined coefficients, where we must make use of the fact that, by definition, the curve defined by the nth-degree polynomial passes through exactly $n + 1$ points on the curve $f(x)$. Inserting any one of the given tabulated points x_i into the polynomial should then give the exact answer $f(x_i)$. Given $n + 1$ such points, we could therefore write the $n + 1$ simultaneous equations

$$p_n(x_0) = a_n x_0^n + a_{n-1} x_0^{n-1} + \cdots + a_1 x_0 + a_0 = f(x_0)$$
$$p_n(x_1) = a_n x_1^n + a_{n-1} x_1^{n-1} + \cdots + a_1 x_1 + a_0 = f(x_1)$$
$$\vdots$$
$$p_n(x_n) = a_n x_n^n + a_{n-1} x_n^{n-1} + \cdots + a_1 x_n + a_0 = f(x_n).$$

$$(8\text{-}3)$$

Since the x_i and the $f(x_i)$ are all given, we have here $n + 1$ equations (linear in the a_i terms) to be solved for the unknowns a_i. (Please note that the situation here is reversed from that of Chapter 5, where the a's were known and the x's were the unknowns.)

Normally, we know that $n + 1$ distinct points do determine a unique nth-degree curve and so we suspect that a unique solution exists for the above set of equations. In terms of our knowledge of linear simultaneous equations, we could examine the determinant of the coefficients of the unknown a_i's, which is

$$\begin{vmatrix} x_0^n & x_0^{n-1} & x_0^{n-2} & \cdots & x_0^2 & x_0 & 1 \\ x_1^n & x_1^{n-1} & x_1^{n-2} & \cdots & x_1^2 & x_1 & 1 \\ x_2^n & x_2^{n-1} & x_2^{n-2} & \cdots & x_2^2 & x_2 & 1 \\ \vdots & & & & & & \vdots \\ x_n^n & x_n^{n-1} & x_n^{n-2} & \cdots & x_n^2 & x_n & 1 \end{vmatrix}.$$

Except for being written backwards (left to right), the above determinant is generally called a *Vandermonde determinant*. It has been studied quite extensively, and it can be shown that it is nonzero as long as all the x_i are distinct. Thus, the system of equations has a solution since its characteristic determinant is nonzero.

We could therefore solve the $n + 1$ equations (8-3) for the $n + 1$ unknowns

a_i and thus have our interpolating polynomial. This is the most straightforward way of finding the nth degree interpolating polynomial, though not the easiest.

8.4 Example 8-1. Use of the Interpolating Polynomial

In this example, we shall start with the following table of sines:

$$\sin\ 0° = 0.0$$
$$\sin 30° = 0.5$$
$$\sin 60° = 0.8660254038$$
$$\sin 90° = 1.0$$

find a third-degree interpolating polynomial, and then use it to provide a table of sines from 0 to 90 degrees in steps of 10 degrees. For the sake of simplicity we shall use degrees instead of radians as the argument.

Using the general third-degree interpolating polynomial

$$p_3(x) = a_3 x^3 + a_2 x^2 + a_1 x + a_0,$$

we use the four points given in the table to write the four simultaneous equations which must all be satisfied:

$$p_3(0) = a_3(\ 0)^3 + a_2(\ 0)^2 + a_1(\ 0) + a_0 = 0.0$$
$$p_3(30) = a_3(30)^3 + a_2(30)^2 + a_1(30) + a_0 = 0.5$$
$$p_3(60) = a_3(60)^3 + a_2(60)^2 + a_1(60) + a_0 = 0.8660254038$$
$$p_3(90) = a_3(90)^3 + a_2(90)^2 + a_1(90) + a_0 = 1.0.$$

Proceeding as in Chapter 5, we construct the augmented matrix

$$\begin{bmatrix} 0 & 0 & 0 & 1.0 & 0.0 \\ 27000 & 900 & 30 & 1.0 & 0.5 \\ 216000 & 3600 & 60 & 1.0 & 0.8660254038 \\ 729000 & 8100 & 90 & 1.0 & 1.0 \end{bmatrix}$$

of the coefficients and the constants to the right of the equals sign. Using the Gauss elimination program of Example 5-3, we find the solution of this system of equations as

$$a_0 = 0.0$$
$$a_1 = 0.17809\ 84095 \times 10^{-1}$$
$$a_2 = -0.19943\ 54803 \times 10^{-4}$$
$$a_3 = -0.60540\ 87040 \times 10^{-6}.$$

The required table of sines is then produced by the following simple program:

```
      A3 =    -0.6054087040   E-06
      A2 =    -0.1994354803   E-04
      A1 =     0.1780984095   E-01
      A0 =     0.0000000000   E 00
      WRITE (3, 650)
  650 FORMAT (3X, 'DEG', 6X, 'INTERPOLATED', 5X, 'TRUE SINE', 11X,
     1    'ERROR')
      DO 700 I = 1, 10
      DEG = (I - 1) * 10
C     TERM  (A3 * DEG**3 + A2 * DEG**2 + A1 * DEG + A0) IS NOW FOUND BY
C        HORNER'S RULE FOR BEST ACCURACY
      CALC = (((A3) * DEC + A2) * DEG + A1) * DEG + A0
      TRUE =  SIN (DEG / 57.2957795)
      ERROR = CALC - TRUE
      WRITE (3, 750) DEG, CALC, TRUE, ERROR
  750 FORMAT (' ', F5.0, 2F16.6, E20.7)
  700 CONTINUE
      CALL  EXIT
      END
```

Running the program (with extended precision) gives us the following table of sines:

DEG	INTERPOLATED	TRUE SINE	ERROR
0.	0.000000	0.000000	0.0000000E 00
10.	0.175498	0.173648	0.1850468E-02
20.	0.343376	0.342020	0.1355986E-02
30.	0.500000	0.500000	-0.4656612E-09
40.	0.641737	0.642787	-0.1049805E-02
50.	0.764957	0.766044	-0.1087354E-02
60.	0.866025	0.866025	0.0000000E 00
70.	0.941310	0.939692	0.1617674E-02
80.	0.987179	0.984807	0.2371558E-02
90.	1.000000	1.000000	-0.9313225E-09

Since the coefficients of the interpolating polynomial were found by assuming that it matched the sine function exactly at 0, 30, 60, and 90 degrees, we would expect zero error at these angles. This is true at 0 and 60 degrees, but only approximately true at 30 and 90 degrees, because of roundoff error as well as error in the FORTRAN SIN function.

Between the tabular values, the error seems to be around 2×10^{-3} or so, and changes sign between the points; this can be seen in Figure 8-5 which shows the sine curve and the interpolating polynomial. At the four tabulated points, the interpolating polynomial exactly matches the true sine; between these points the polynomial is in error.

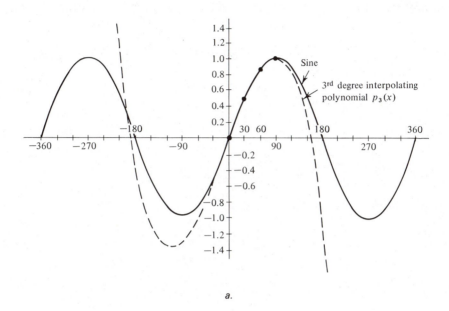

a.

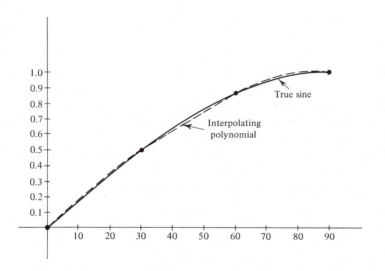

b. Expanded view between 0 and 90 degrees.

Figure 8-5. Sine curve and third-degree interpolating polynomial.

8.5 The Lagrange Interpolating Polynomial

The interpolating polynomial form used above is not the easiest to apply, since it often requires the solution of a large system of simultaneous equations. There is another way of writing the same polynomial, called the *Lagrange* form.

Let us derive the first-degree Lagrange polynomial. As before, we seek a polynomial of the form

$$p_1(x) = a_1 x + a_0,$$

such that $p_1(x_0) = f(x_0)$, and $p_1(x_1) = f(x_1)$, at the two tabulated points x_0 and x_1.

We write these three equations as follows:

$$p_1(x) - a_1 x - a_0 = 0$$
$$f(x_0) - a_1 x_0 - a_0 = 0$$
$$f(x_1) - a_1 x_1 - a_0 = 0.$$

Furthermore, let us define a constant $c = 1$, so that

$$cp_1(x) - a_1 x - a_0 = 0$$
$$cf(x_0) - a_1 x_0 - a_0 = 0$$
$$cf(x_1) - a_1 x_1 - a_0 = 0,$$

where we know that $c = 1$, and that a_1 and a_0 exist and are unique. Now, using the terminology of Chapter 5, we can treat these three equations as being three homogeneous equations in the three " unknowns " c, a_1, and a_0. We know that a set of homogeneous equations has a nonzero solution only if the determinant of the system is zero. The determinant of the coefficients is therefore

$$\begin{vmatrix} p_1(x) & -x & -1 \\ f(x_0) & -x_0 & -1 \\ f(x_1) & -x_1 & -1 \end{vmatrix} = 0.$$

If we now expand the determinant by the first column, we have

$$p_1(x)(x_0 - x_1) - f(x_0)(x - x_1) + f(x_1)(x - x_0) = 0.$$

Solving for $p_1(x)$ we have

$$p_1(x) = f(x_0)\frac{(x - x_1)}{(x_0 - x_1)} + f(x_1)\frac{(x - x_0)}{(x_1 - x_0)},$$

which is the first-degree Lagrange interpolating polynomial. The expression is more generally written as

$$p_1(x) = f(x_0)L_0(x) + f(x_1)L_1(x),$$

with the L's defined separately as

$$L_0(x) = \frac{(x - x_1)}{(x_0 - x_1)}, \qquad L_1(x) = \frac{(x - x_0)}{(x_1 - x_0)}.$$

In general, the nth-degree Lagrange interpolating polynomial is written in the same way as

$$p_n(x) = f(x_0)L_0(x) + f(x_1)L_1(x) + \cdots + f(x_n)L_n(x). \qquad (8\text{-}4)$$

Since we require that at the tabulated points x_i we must have

$$p_n(x_i) \equiv f(x_i),$$

exactly, we could arrange this if the L's were chosen so that at x_i the $L_i(x)$ term equals 1, while all the other L's are equal to zero. Then, for example, at x_1 we would have

$$p_n(x_1) = f(x_0)\cdot 0 + f(x_1)\cdot 1 + f(x_2)\cdot 0 + \cdots + f(x_n)\cdot 0$$
$$= f(x_1).$$

This is easily arranged if we pick the L's so that

$$L_i(x) = \frac{(x - x_0)(x - x_1) \cdots (x - x_{i-1})(x - x_{i+1}) \cdots (x - x_n)}{(x_i - x_0)(x_i - x_1) \cdots (x_i - x_{i-1})(x_i - x_{i+1}) \cdots (x_i - x_n)}.$$
$$(8\text{-}5)$$

At any x_j, such that $x_j \neq x_i$, one of the factors in the numerator is zero, and $L_i(x)$ is zero. At $x = x_i$, on the other hand, the numerator and denominator are exactly equal, and $L_i(x)$ equals 1. Since the $(x - x_i)$ term is missing from the expression for $L_i(x)$, its numerator is an nth degree polynomial in x; thus, since x does not appear in the denominator, each $L_i(x)$ itself is an nth degree polynomial in x, and so is $p_n(x)$.

Using the product sign $\prod$ we can express the L's in the more compact form

$$L_i(x) = \frac{\displaystyle\prod_{\substack{j=0 \\ j \ne i}}^{n} (x - x_j)}{\displaystyle\prod_{\substack{k=0 \\ k \ne i}}^{n} (x_i - x_k)} = \prod_{\substack{j=0 \\ j \ne i}}^{n} \frac{(x - x_j)}{(x_i - x_j)}. \tag{8-6}$$

8.6 Example 8-2. Lagrange Interpolating Polynomial Method

As in Example 8-1, we again use an interpolating polynomial to produce a table of sines by interpolating for given points of 0°, 30°, 60°, and 90°.

In the following program we define X0 through X3 and FX0 through FX3 as the given points, and EL0 through EL3 as the Lagrange coefficients $L_0(x)$ through $L_3(x)$; the name EL is used instead of L to avoid having to declare L as real rather than integer.

Since only four coefficients EL are needed, it is easy to express them explicitly as, for example,

$$\text{EL0} = L_0(x) = \frac{(x - x_1)(x - x_2)(x - x_3)}{(x_0 - x_1)(x_0 - x_2)(x_0 - x_3)},$$

and so on; if more points had been given so that more $L_i(x)$ were required, then some form of DO loop to calculate these terms would have been preferable.

The completed Lagrange interpolating polynomial program is the following:

```
      X0 = 0.0
      X1 = 30.0
      X2 = 60.0
      X3 = 90.0
      FX0 = 0.0
      FX1 = 0.5
      FX2 = 0.8660254038
      FX3 = 1.0
      WRITE (3, 650)
  650 FORMAT (3X, 'DEG', 6X, 'INTERPOLATED', 5X, 'TRUE SINE', 11X,
     1 'ERROR')
      DO 700 I = 1, 10
      X   = (I - 1) * 10
      EL0 = ((X-X1)*(X-X2)*(X-X3)) / ((X0-X1)*(X0-X2)*(X0-X3))
      EL1 = ((X-X0)*(X-X2)*(X-X3)) / ((X1-X0)*(X1-X2)*(X1-X3))
      EL2 = ((X-X0)*(X-X1)*(X-X3)) / ((X2-X0)*(X2-X1)*(X2-X3))
      EL3 = ((X-X0)*(X-X1)*(X-X2)) / ((X3-X0)*(X3-X1)*(X3-X2))
```

```
      CALC = FX0*EL0 + FX1*EL1 + FX2*EL2 + FX3*EL3
      TRUE =  SIN ( X  / 57.2957795)
      ERROR = CALC - TRUE
      WRITE (3, 750)   X, CALC, TRUE, ERROR
750 FORMAT (' ', F5.0, 2F16.6, E20.7)
700 CONTINUE
      CALL  EXIT
      END
```

We then obtain the following results:

DEG	INTERPOLATED	TRUE SINE	ERROR
0.	0.000000	0.000000	0.0000000E 00
10.	0.175498	0.173648	0.1850468E-02
20.	0.343376	0.342020	0.1355986E-02
30.	0.500000	0.500000	-0.4656612E-09
40.	0.641737	0.642787	-0.1049805E-02
50.	0.764957	0.766044	-0.1087354E-02
60.	0.866025	0.866025	0.0000000E 00
70.	0.941310	0.939692	0.1617673E-02
80.	0.987179	0.984807	0.2371558E-02
90.	1.000000	1.000000	-0.1862645E-08

Comparing these results with those of Example 8-1, we note that the interpolated values are the same to within six decimal digits, although the errors differ slightly. This is understandable, since the roundoff error should be different if the same calculations are done in a different order.

8.7 Uniqueness of the Interpolating Polynomial

We have so far presented two different ways of determining an interpolating polynomial of degree n, and there are still others. We may therefore wonder whether these methods all give the same polynomial (in different forms) or whether the polynomials are different. It is easy to show that they must all be the same.

As proof, consider a tabulated set of $n + 1$ points, to which we interpolate with an nth-degree polynomial. Suppose that we obtain the nth-degree polynomial $P_n(x)$ using one method, and some other polynomial $Q_n(x)$ using some other method; each of these is of degree n with x^n being the highest power.

Noting that there are $n + 1$ tabulated points which are exactly matched by both polynomials, let us get their difference

$$R_n(x) = P_n(x) - Q_n(x).$$

$R_n(x)$ is a polynomial of degree n or less. There are $n + 1$ values of x, however, where P_n and Q_n have the same value, so that R_n is consequently zero at those $n + 1$ points.

An nth-degree polynomial, however, can have only n zeroes, not $n + 1$

zeroes. Hence $R_n(x)$ cannot exist *unless* it is identically zero for *any* x. But if this is so, then $P_n(x)$ and $Q_n(x)$ must both be the same. Thus the interpolating polynomial is unique.

As a corollary, suppose that the tabulated function $f(x)$ itself is an nth-degree polynomial which we attempt to approximate by an nth-degree polynomial $P_n(x)$. If these two polynomials are to coincide at $n + 1$ points, they too must be the same. This just says that we can reconstruct $f(x)$ from a table of $n + 1$ points.

8.8 Aitken's Interpolating Polynomial

Having shown that all interpolating polynomials are merely different forms of the same polynomial, we seek some particularly easy way of writing the polynomial in a manner which makes it easy to use. Aitken's form is one such way.

Let us return to the first-degree Lagrange interpolating polynomial

$$p_1(x) = f(x_0)\frac{(x - x_1)}{(x_0 - x_1)} + f(x_1)\frac{(x - x_0)}{(x_1 - x_0)},$$

which can be rewritten as

$$p_1(x) = \frac{1}{(x_1 - x_0)}[f(x_0)(x_1 - x) - f(x_1)(x_0 - x)].$$

The quantity inside the brackets can be rewritten as a determinant:

$$p_1(x) = \frac{1}{(x_1 - x_0)}\begin{vmatrix} f(x_0) & (x_0 - x) \\ f(x_1) & (x_1 - x) \end{vmatrix}$$

which provides a fairly convenient way of writing the first-degree polynomial. This particular polynomial exactly matches the tabulated function at x_0 and x_1; to indicate this more clearly, we shall change the notation a bit and write this as p_{01}, where the subscript $_{01}$ indicates that a perfect fit is obtained at x_0 and x_1. We now have

$$p_{01}(x) = \frac{1}{(x_1 - x_0)}\begin{vmatrix} f(x_0) & (x_0 - x) \\ f(x_1) & (x_1 - x) \end{vmatrix}.$$

Using the same notation, let us next write a second interpolating polynomial $p_{02}(x)$ which provides a perfect fit at x_0 and at x_2; it could be used for interpolating in the range between x_0 and x_2:

$$p_{02}(x) = \frac{1}{(x_2 - x_0)} \begin{vmatrix} f(x_0) & (x_0 - x) \\ f(x_2) & (x_2 - x) \end{vmatrix}.$$

In general, we could use x_0 and any other tabulated point, such as the one at x_i, to get a polynomial useful for interpolating between x_0 and x_i:

$$p_{0i}(x) = \frac{1}{(x_i - x_0)} \begin{vmatrix} f(x_0) & (x_0 - x) \\ f(x_i) & (x_i - x) \end{vmatrix}.$$

Depending on the particular value of x for which we wish to interpolate, the different p_{0i} will in general give different interpolated values, which will have varying degrees of error.

Let us now combine p_{01} and p_{02} in the following way to give a new polynomial which we shall call p_{012}:

$$p_{012}(x) = \frac{1}{(x_2 - x_1)}[p_{01}(x)(x_2 - x) - p_{02}(x)(x_1 - x)],$$

which can also be written as

$$p_{012}(x) = \frac{1}{(x_2 - x_1)} \begin{vmatrix} p_{01}(x) & (x_1 - x) \\ p_{02}(x) & (x_2 - x) \end{vmatrix}. \tag{8-7}$$

Hence p_{012} could be interpreted as an interpolation between p_{01} and p_{02}.

Now, since $p_{01}(x)$ and $p_{02}(x)$ are each of first degree, p_{012} is of second degree since it has another factor of x. Moreover, it gives exact values of $f(x)$ at the tabulated points x_0, x_1, and x_2, as we can see from the following. Rewriting (8-7), we have:

$$p_{012}(x) = \frac{1}{(x_2 - x_1)}[p_{01}(x)(x_2 - x) - p_{02}(x)(x_1 - x)].$$

At $x = x_0$, $p_{01}(x_0)$ and $p_{02}(x_0)$ both equal $f(x_0)$, and we have

$$p_{012}(x_0) = \frac{1}{(x_2 - x_1)}[f(x_0)(x_2 - x_0) - f(x_0)(x_1 - x_0)]$$

$$= \frac{1}{(x_2 - x_1)}[f(x_0)(x_2 - x_1)] = f(x_0).$$

At $x = x_1$, $p_{01}(x_1)$ equals $f(x_1)$, and $(x_1 - x)$ is zero:

$$p_{012}(x_1) = \frac{1}{(x_2 - x_1)}[f(x_1)(x_2 - x_1) - p_{02}(x_1)(0)]$$

$$= f(x_1).$$

Finally, at $x = x_2$, $p_{02}(x_2)$ equals $f(x_2)$, and $(x_2 - x)$ is zero:

$$p_{012}(x_2) = \frac{1}{(x_2 - x_1)}[p_{01}(x_2)(0) - f(x_2)(x_1 - x_2)]$$

$$= f(x_2).$$

Hence $p_{012}(x)$ is a second-degree polynomial, and it also matches $f(x)$ exactly at the three tabulated points at x_0, x_1, and x_2. But by the results of the previous section, we know that the interpolating polynomial is unique, so that p_{012} must be *the* second-degree interpolating polynomial.

In the same way as we got $p_{012}(x)$, we could have defined a $p_{01i}(x)$ which would have interpolated exactly at the three tabulated points x_0, x_1, and x_i: in terms of the determinant notation used earlier, we would have written p_{01i} as follows:

$$p_{01i}(x) = \frac{1}{(x_i - x_1)}\begin{vmatrix} p_{01}(x) & (x_1 - x) \\ p_{0i}(x) & (x_i - x) \end{vmatrix}. \tag{8-8}$$

Each of these p_{01i} would have been a second-degree interpolating polynomial, obtained in a rather painless and systematic way.

But we need not stop here. We can show easily that the expression

$$p_{0123}(x) = \frac{1}{(x_3 - x_2)}\begin{vmatrix} p_{012}(x) & (x_2 - x) \\ p_{013}(x) & (x_3 - x) \end{vmatrix} \tag{8-9}$$

is in turn a third-degree interpolating polynomial which exactly interpolates at the four tabulated points at x_0, x_1, x_2, and x_3. Similarly, we could derive p_{012i} for any of the other x_i. By carrying out this process further we could develop the interpolating polynomial to as high a degree as we wish. To make sure we develop the process in an orderly way, we usually proceed in the following order:

1. Using the tabulated points x_0 through x_n, we first calculate p_{01} through p_{0n}.
2. Next we calculate p_{012} through p_{01n}.
3. Then we calculate p_{0123} through p_{012n}.
4. We continue with p_{01234} through p_{0123n}, then p_{012345} through p_{01234n}, and so on.

This process is usually written in the form of a table as follows:

Given Values		*Calculated Values*					
x_i $f(x_i)$		p_{0i}	p_{01i}	p_{012i}	p_{0123i}	p_{01234i}	$\cdots$
x_0 $f(x_0)$							
x_1 $f(x_1)$		p_{01}					
x_2 $f(x_2)$		p_{02}	p_{012}				
x_3 $f(x_3)$		p_{03}	p_{013}	p_{0123}			
x_4 $f(x_4)$		p_{04}	p_{014}	p_{0124}	p_{01234}		
x_5 $f(x_5)$		p_{05}	p_{015}	p_{0125}	p_{01235}	p_{012345}	
$\vdots$							
x_i $f(x_i)$		p_{0i}	p_{01i}	p_{012i}	p_{0123i}	p_{01234i}	$\cdots$
$\vdots$							
x_n $f(x_n)$		p_{0n}	p_{01n}	p_{012n}	p_{0123n}	p_{01234n}	$\cdots$

In this table, the first two columns represent the given tabulated points. The calculated values are found one column at a time, from top to bottom. Each value of $p_{01\ldots i}$ is found from (1) the p immediately to its left, and (2) the p at the top of the column to the left. For example, p_{0123i} is calculated from p_{012i} and p_{0123}, using the equation

$$p_{0123i}(x) = \frac{1}{(x_i - x_3)}\begin{vmatrix} p_{0123}(x) & (x_3 - x) \\ p_{012i}(x) & (x_i - x) \end{vmatrix}.$$

In general, we might find any $p_{012\ldots h, i}$ from the following expression:

$$p_{012\ldots h, i}(x) = \frac{1}{(x_i - x_h)}\begin{vmatrix} p_{012\ldots h}(x) & (x_h - x) \\ p_{012\ldots h-1, i}(x) & (x_i - x) \end{vmatrix}. \tag{8-10}$$

Each succeeding column of the table has one less value to calculate, until eventually we reach the last column which has only one value; that is the answer.

To illustrate the method, we take a table of values such as

i	x_i	$f(x_i)$
0	0	0
1	1	1
2	3	27
3	5	125
4	6	216

The function is obviously $f(x) = x^3$, so at most a third-degree interpolating polynomial should be required to give an interpolated value at $x = 2$ exactly. The resulting table of p's (each evaluated at $x = 2$) is

x_i	$f(x_1)$	p_{0i}	p_{01i}	p_{012i}	p_{0123i}
0	0				
1	1	$p_{01} = 2$			
3	27	$p_{02} = 18$	$p_{012} = 10$		
5	125	$p_{03} = 50$	$p_{013} = 14$	$p_{0123} = 8$	
6	216	$p_{04} = 72$	$p_{014} = 16$	$p_{0124} = 8$	$p_{01234} = 8$

Since $f(x)$ was a cubic equation, only a third-degree polynomial was required, and the correct answer of $f(2) = 2^3 = 8$ was obtained quite soon as p_{0123}. In general, however, the number of points given in a table is less than the degree of the function, so that we will never reach the exact answer by interpolation. Yet, the diagonal terms $p_{01}, p_{012}, p_{0123}, p_{01234}$, etc. will generally approach the correct interpolated value more closely; by letting the computer program monitor the difference between successive interpolations, we can interpolate until the difference between successive interpolations becomes arbitrarily small.

In other words, Aitken's method has one big advantage: we need not decide in advance what degree polynomial to take, but can let the program decide when to stop. Theoretically, this is a great advantage; practically, we may find that the process converges to the wrong value simply because roundoff error accumulates and makes the higher-degree interpolations less accurate.

8.9 Error of the Interpolating Polynomial

We may develop an expression for the error between the interpolated value $p_n(x)$ and the actual value $f(x)$. This calculated error represents the error of the polynomial itself, and ignores roundoff errors which vary, depending on the method.

Intuitively, we suspect that a very smooth function can be easily interpolated with a low-degree interpolating polynomial, while a function with many bends and corners would need a high-degree interpolation. In fact, suppose that the function $f(x)$ is itself an nth-degree polynomial, so that its $(n + 1)$th derivative is zero. If we approximate this function with an nth-degree interpolating polynomial $p_n(x)$, we find that the interpolating polynomial exactly matches the function and the error is zero. If, on the other hand, $f(x)$ is of higher than nth degree, so that its $(n + 1)$th derivative is not zero, then we do have an error. We therefore think, quite rightly, that the error of an nth-degree interpolating polynomial is somehow dependent on the $(n + 1)$th derivative of $f(x)$, if that derivative exists.

To derive an expression for the error, let us consider a continuous function $f(x)$ which has at least the first $n + 1$ derivatives. We have the function tabulated for $n + 1$ points $x_0, x_1, \ldots, x_n$, through which we pass an nth-degree interpolating polynomial $p_n(x)$. To find the value of the function at any desired point $\bar{x}$ we calculate $p_n(\bar{x})$. We will now prove that for any $\bar{x}$ in the region containing the tabulated points x_i we have an error $\varepsilon_n(\bar{x})$ which equals

$$\varepsilon_n(\bar{x}) = f(\bar{x}) - p_n(\bar{x}) = f^{(n+1)}(\xi)\frac{(\bar{x} - x_0)(\bar{x} - x_1) \cdots (\bar{x} - x_n)}{(n + 1)!},$$

where $f^{(n+1)}(\xi)$ is the $(n + 1)$th derivative of $f(x)$ evaluated at some (unknown) point $x = \xi$ in the interval containing the tabulated points x_i and the desired point $\bar{x}$.

To simplify the notation, let us introduce a new function $\phi(x)$

$$\phi(x) = (x - x_0)(x - x_1) \cdots (x - x_n), \tag{8-11}$$

so that we can write

$$\varepsilon_n(\bar{x}) = f(\bar{x}) - p_n(\bar{x}) = f^{(n+1)}(\xi)\frac{\phi(\bar{x})}{(n + 1)!}.$$

Now let us define a new function $e(x)$ such that

$$e(x) = f(x) - p_n(x) - \phi(x)\frac{f(\bar{x}) - p_n(\bar{x})}{\phi(\bar{x})}.$$

In this latter equation, x is any value of x in the interval which contains the tabulated points, while $\bar{x}$ is the particular value of x where we wish to evaluate the function. Thus, having decided where to evaluate the function, we can look upon x as a variable, while $\bar{x}$ is a constant.

The new function $e(x)$ has some interesting properties. At the tabulated points x_i, we note that $\phi(x_i) = 0$ since one term in $\phi(x_i)$ will then be zero; also $p_n(x_i) = f(x_i)$, so that

$$e(x_i) = f(x_i) - f(x_i) - 0\frac{f(\bar{x}) - p_n(\bar{x})}{\phi(\bar{x})}$$

$$= 0.$$

If, on the other hand, $x = \bar{x}$, then we have

$$e(\bar{x}) = f(\bar{x}) - p_n(\bar{x}) - \phi(\bar{x})\frac{f(\bar{x}) - p_n(\bar{x})}{\phi(\bar{x})}$$

$$= 0.$$

Hence there is a total of $n + 2$ values of x in the interval where $e(x) = 0$.

Now we note that $\phi(\bar{x})$ is not zero, since presumably $\bar{x} \neq x_i$, one of the tabulated points; if it were, we would not need to interpolate at all, but could simply use the tabulated point x_i. Hence if $\phi(\bar{x})$ is nonzero, and each of the other components of $e(x)$ is either a constant or else a continuous and differentiable function, then $e(x)$ itself is continuous and differentiable and we can apply the mean value theorem to it.

The mean value theorem (see Chapter 3) states that if a function $f(x)$ is continuous and has a continuous derivative, then between any two points at $x = a$ and $x = b$ on the curve there exists a point at $x = \xi$ where the slope of the tangent to the curve is the same as the slope of the secant between $(a, f(a))$ and $(b, f(b))$. Applying this theorem to $e(x)$, we see that between any two zeroes of $e(x)$ there must be at least one point where the slope of the tangent is zero; hence the derivative is zero. This result is generally known as *Rolle's Theorem*. Hence if $e(x)$ has $n + 2$ zeroes on the interval containing the tabulated points, then the first derivative $e'(x)$ has at least $n + 1$ zeroes. Applying the theorem again, we see that $e''(x)$ must have at least n zeroes, and so on, until $e^{(n+1)}(x)$ must have at least one zero. Suppose then that such a zero occurs at $x = \xi$.

Let us now take $e(x)$ and formally differentiate it $n + 1$ times:

$$e^{(n+1)}(x) = \frac{d^{n+1}}{dx^{n+1}}e(x)$$

$$= \frac{d^{n+1}}{dx^{n+1}}\left\{ f(x) - p_n(x) - \phi(x)\frac{f(\bar{x}) - p_n(\bar{x})}{\phi(\bar{x})} \right\}.$$

Upon differentiating, we note that $p_n(x)$ is an nth-degree polynomial so that its $(n + 1)$th derivative is zero; $\phi(x)$ is an $(n + 1)$th-degree polynomial

$$\phi(x) = (x - x_0)(x - x_1)(x - x_2) \cdots (x - x_n) = x^{n+1}$$
$$+ \text{ lower degree terms,}$$

whose $(n + 1)$th derivative is $(n + 1)!$; finally, the terms in $\bar{x}$ are constants. Hence we get

$$e^{(n+1)}(x) = f^{(n+1)}(x) - (n + 1)!\frac{f(\bar{x}) - p_n(\bar{x})}{\phi(\bar{x})}.$$

Evaluating at $x = \xi$ we should get zero:

$$e^{(n+1)}(\xi) = f^{(n+1)}(\xi) - (n+1)!\frac{f(\bar{x}) - p_n(\bar{x})}{\phi(\bar{x})} = 0,$$

where the equation can be solved for

$$\varepsilon_n(\bar{x}) = f(\bar{x}) - p_n(\bar{x}) = f^{(n+1)}(\xi)\frac{\phi(\bar{x})}{(n+1)!}.$$

In the resulting error expression, the particular value of ξ depends, of course, on the particular point $\bar{x}$ where an interpolated value is desired; in general we have no idea of just how to pick the right ξ for the error calculation. This problem is analogous to estimating the error of a Taylor series by examination of the remainder term.

If, on the other hand, we could find some upper bound for $f^{(n+1)}(\xi)$ for any ξ in the interval containing the tabulated points x_i and the desired point $\bar{x}$, then we could get an upper bound on the error. Thus, suppose that for any ξ in the interval we have

$$\left|f^{(n+1)}(\xi)\right| \leqslant M;$$

then we can write

$$\left|\varepsilon_n(\bar{x})\right| \leqslant \left|M\frac{\phi(\bar{x})}{(n+1)!}\right|.$$

where $\phi(x)$ is defined in Eq. (8-11).

As an example, let us obtain an error estimate for the error in the interpolated sine of 10 degrees as obtained in example 8-1. To use the maximum error equation, we need the maximum value of $f^{\mathrm{IV}}(\xi)$ in the range from 0 to 90 degrees. The fourth derivative of the sine is again the sine, but here we must be careful since the interpolation is done in degrees, not in radians. We may write

$$f(x) = \sin(x/57.2957),$$

where x is in degrees and the division by 57.2957 (which is equal to $180/\pi$) is done to convert to radians. We then differentiate four times as follows:

$$d(\sin u) = \cos u \, du$$
$$f'(x) = \cos(x/57.2957)/57.2957,$$

and similarly

$$f''(x) = -\sin(x/57.2957)/(57.2957)^2$$
$$f'''(x) = -\cos(x/57.2957)/(57.2957)^3$$
$$f^{IV}(x) = \sin(x/57.2957)/(57.2957)^4$$
$$\leqslant 1/(57.2957)^4.$$

Hence the error expression at $\bar{x} = 10$ degrees is

$$|\varepsilon_n(10)| \leqslant \left| \frac{1}{(57.2957)^4} \frac{(10-0)(10-30)(10-60)(10-90)}{4!} \right|$$

$$\leqslant \left| \frac{1}{1.08 \times 10^7} \frac{(10)(-20)(-50)(-80)}{4!} \right|$$

$$\leqslant \left| \frac{-8 \times 10^5}{24 \times 1.08 \times 10^7} \right| = 0.31 \times 10^{-2}.$$

The actual error in Example 8-1 is 0.185×10^{-2}, which is less than the maximum error calculated above. In general, the maximum error estimate may be an overly conservative value, but it is the best we can do.

8.10 Difference Calculus

The interpolating polynomial as we know it so far can be applied to tabular points $x_0, x_1, \ldots, x_n$ which need not necessarily be evenly spaced. If the tabulated points are spaced evenly, however, different and sometimes more convenient methods can be used; these are based on the so-called *calculus of finite differences*.

As an introductory example, let us consider the following table of values for the function $f(x) = x^3 + 5$:

x_i	$f(x_i)$
0	5
1	6
2	13
3	32
4	69
5	130
6	221

As in this example, the x_i must be evenly spaced by some increment h so that

$$x_i = x_{i-1} + h = x_{i-2} + 2h = x_{i-3} + 3h = \cdots,$$

and, most important,

$$x_i = x_0 + ih.$$

Suppose that we now add a third column to the above table, such that the third column will consist of the differences between the $f(x_i)$ in the second column:

x_i	$f(x_i)$	$f(x_{i+1}) - f(x_i)$
0	5	
		$6 - 5 = 1$
1	6	
		$13 - 6 = 7$
2	13	
		$32 - 13 = 19$
3	32	
		$69 - 32 = 37$
4	69	
		$130 - 69 = 61$
5	130	
		$221 - 130 = 91$
6	221	

In the same way, let us add additional columns such that each column gives the differences between entries in the previous column:

x_i	$f(x_i)$	First difference	Second difference	Third difference	Fourth difference
0	5				
		1			
1	6		6		
		7		6	
2	13		12		0
		19		6	
3	32		18		0
		37		6	
4	69		24		0
		61		6	
5	130		30		
		91			
6	221				

There are several interesting characteristics of this table; suppose we differentiate $f(x)$:

$$f(x) = x^3 + 5,$$
$$f'(x) = 3x^2,$$
$$f''(x) = 6x,$$
$$f'''(x) = 6,$$
$$f^{\text{IV}}(x) = 0.$$

We note that the third derivative equals 6, and so do the third differences; similarly, the fourth derivative and the fourth differences are both zero. We therefore suspect that the differences must somehow be related to derivatives. This is quite correct, and will be useful in our later work.

But first, to put the theory on a better footing, let us examine differences on a more theoretical basis. First, let us define a *forward difference operator* Δ as follows:

Definition *The* forward difference operator Δ *denotes the operation performed by the right-hand side of the following:*

$$\Delta f(x) = f(x + h) - f(x),$$

where h is some pre-determined constant.

If, for example, $y = f(x)$, then we are all familiar with the notation

$$\Delta y = f(x + h) - f(x) = \Delta f(x).$$

In this way, the difference operator Δ is already well known. For example, we could write

$$\Delta \cos x = \cos (x + h) - \cos x$$
$$\Delta \sin x = \sin (x + h) - \sin x$$
$$\Delta \log x = \log (x + h) - \log x,$$

and so on.

Now, suppose that we have a table of values of some function $f(x)$, with the x_i evenly spaced so that we can write

$$x_1 = x_0 + h$$
$$x_2 = x_1 + h$$
$$x_3 = x_2 + h$$
$$\vdots$$

then we could write

$$\Delta f(x_0) = f(x_0 + h) - f(x_0)$$
$$= f(x_1) - f(x_0),$$

or, in the case of our table of $x^3 + 5$, we would have

$$\Delta f(x_0) = 6 - 5 = 1.$$

In the same way, each of the other terms in the first differences column of the table could be defined in terms of the Δ operator. The second term in that column is, for example,

$$\Delta f(x_1) = f(x_1 + h) - f(x_1)$$
$$= f(x_2) - f(x_1)$$
$$= 13 - 6 = 7.$$

Continuing down the first difference column, each of the terms can be written as a $\Delta f(x_i)$.

But we can define higher "powers" of Δ as well. Thus if

$$\Delta f(x) = f(x + h) - f(x),$$

then we could write

$$\Delta^2 f(x) = \Delta[\Delta f(x)]$$
$$= \Delta[f(x + h) - f(x)]$$
$$= [\Delta f(x + h)] - [\Delta f(x)]$$
$$= [f(x + h + h) - f(x + h)] - [f(x + h) - f(x)]$$
$$= f(x + 2h) - 2f(x + h) + f(x).$$

For example, we may look for

$$\Delta^2 f(x_0) = f(x_0 + 2h) - 2f(x_0 + h) + f(x)$$
$$= f(x_2) - 2f(x_1) + f(x)$$
$$= 13 - 2 \cdot 6 + 5$$
$$= 13 - 12 + 5$$
$$= 6,$$

which is just the top term in the second difference column. Using the difference operator Δ, then, we could express *any* term in the table of differences in terms of Δ. This would give us the following table:

x_i	$f(x_i)$	First difference	Second difference	Third difference	Fourth difference
x_0	$f(x_0)$				
		$\Delta f(x_0)$			
x_1	$f(x_1)$		$\Delta^2 f(x_0)$		
		$\Delta f(x_1)$		$\Delta^3 f(x_0)$	
x_2	$f(x_2)$		$\Delta^2 f(x_1)$		$\Delta^4 f(x_0)$
		$\Delta f(x_2)$		$\Delta^3 f(x_1)$	
x_3	$f(x_3)$		$\Delta^2 f(x_2)$		$\Delta^4 f(x_1)$
		$\Delta f(x_3)$		$\Delta^3 f(x_2)$	
x_4	$f(x_4)$		$\Delta^2 f(x_3)$		$\Delta^4 f(x_2)$
		$\Delta f(x_4)$		$\Delta^3 f(x_3)$	$\cdots$
x_5	$f(x_5)$		$\Delta^2 f(x_4)$		
		$\Delta f(x_5)$		$\vdots$	$\vdots$
x_6	$f(x_6)$	$\vdots$	$\vdots$		
$\vdots$	$\vdots$				

As already pointed out, there are great similarities between derivatives and differences. For example,

$$\frac{\Delta f(x)}{h} = \frac{f(x + h) - f(x)}{h}.$$

The derivative $f'(x)$ is defined as the limit of this quantity as h approaches zero. Alternatively, if $f(x)$ is analytic, then the mean value theorem tells us that there exists a ξ between x and $x + h$ such that

$$f'(\xi) = \frac{f(x + h) - f(x)}{h},$$

so that we can write

$$\Delta f(x) = f(x + h) - f(x) = hf'(\xi). \tag{8-12}$$

Assuming a fixed h, this relation tells us that if $f(x)$ is such that $f'(x)$ is a constant or zero, then $\Delta f(x)$ is also a constant, or zero, respectively. This concept can be extended to higher orders of derivatives and differences.

8.11 Use of Differences to Check Table Accuracy

Before we continue to the actual use of differences in interpolation, let us show still another use of a table of differences. Suppose we construct a table of differences for the sines of angles from 0 to 90 degrees in steps of 10 degrees, using a five-place sines table as the starting point:

x_i (degrees)	$f(x_i)$ = $\sin x_i$	Δ	Δ^2	Δ^3	Δ^4	Δ^5	Δ^6
0	0.00000						
		17365					
10	0.17365		−528				
		16837		−511			
20	0.34202		−1039		31		
		15798		−480		14	
30	0.50000		−1519		45		+4
		14279		−435		18	
40	0.64279		−1954		63		−16
		12325		−372		2	
50	0.76604		−2326		65		+19
		9999		−307		21	
60	0.86603		−2633		86		−25
		7366		−221		−4	
70	0.93969		−2854		82		
		4512		−139			
80	0.98481		−2993				
		1519					
90	1.00000						

For the sake of convenience, we have omitted the decimal point and initial zeroes in this table to avoid needless repetition.

Remembering that there is great similarity between differences and derivatives, we note that the first derivative of the sine is the cosine, the second is the negative sine, and so on, in the following sequence:

$$\cos$$
$$-\sin$$
$$-\cos$$
$$\sin$$
$$\cos$$
$$-\sin$$
$$\vdots$$

This helps to explain why the first differences are all positive, the second and third are negative, and the fourth and fifth are (mostly) positive. But the sixth differences alternate signs, whereas the sixth derivative $(-\sin)$ is negative; how can we explain this difference?

This strange behavior can easily be explained when we remember that the values of sines are obtained from a five-place table, and are not accurate to more than five places.

Suppose we write the following table, and deliberately introduce an error term ε into one of the $f(x_i)$, say $f(x_5)$:[4]

x_i $f(x_i)$	Δ	Δ^2	Δ^3	Δ^4	Δ^5	Δ^6
x_0 $f(x_0)$						
	Δf_0					
x_1 $f(x_1)$		$\Delta^2 f_0$				
	Δf_1		$\Delta^3 f_0$			
x_2 $f(x_2)$		$\Delta^2 f_1$		$\Delta^4 f_0$		
	Δf_2		$\Delta^3 f_1$		$\Delta^5 f_0 + \varepsilon$	
x_3 $f(x_3)$		$\Delta^2 f_2$		$\Delta^4 f_1 + \varepsilon$		$\Delta^6 f_0 - 6\varepsilon$
	Δf_3		$\Delta^3 f_2 + \varepsilon$		$\Delta^5 f_1 - 5\varepsilon$	
x_4 $f(x_4)$		$\Delta^2 f_3 + \varepsilon$		$\Delta^4 f_2 - 4\varepsilon$		$\Delta^6 f_1 + 15\varepsilon$
	$\Delta f_4 + \varepsilon$		$\Delta^3 f_3 - 3\varepsilon$		$\Delta^5 f_2 + 10\varepsilon$	
x_5 $f(x_5) + \varepsilon$		$\Delta^2 f_4 - 2\varepsilon$		$\Delta^4 f_3 + 6\varepsilon$		$\Delta^6 f_2 - 20\varepsilon$
	$\Delta f_5 - \varepsilon$		$\Delta^3 f_4 + 3\varepsilon$		$\Delta^5 f_3 - 10\varepsilon$	
x_6 $f(x_6)$		$\Delta^2 f_5 + \varepsilon$		$\Delta^4 f_4 - 4\varepsilon$		$\Delta^6 f_3 + 15\varepsilon$
	Δf_6		$\Delta^3 f_5 - \varepsilon$		$\Delta^5 f_4 + 5\varepsilon$	
x_7 $f(x_7)$		$\Delta^2 f_6$		$\Delta^4 f_5 + \varepsilon$		
	Δf_7		$\Delta^3 f_6$			
x_8 $f(x_8)$		$\Delta^2 f_7$				
	Δf_8					
x_9 $f(x_9)$						

A small error ε propagates to the right and spreads up and down, until it becomes rather large in the higher differences. Moreover, the sign of the error in the higher differences alternates from $+$ to $-$, just as the error in the sines table did. We are therefore led to the conclusion that the alternating signs in the sixth difference column of the sines table were actually due to small errors in the tabulated values; this is indeed true, since the table is only accurate to five digits, so that the error in any value could be as large as ± 0.000005. The resulting error in each value then propagates to the right, and causes alternating signs in the higher differences. Thus the difference table is a handy method for testing the accuracy of a tabulated function.

Knowing how the error propagates from left to right, we may be tempted to use this knowledge to estimate the error in the tabulated values themselves.

---

[4] To shorten the table, we write f_i instead of the clumsier $f(x_i)$.

Generally, this is a dangerous procedure, since we do not know which tabulated value has an error, or how many errors there are. Several of the $f(x_i)$ may have an error term ε_i; by the time these several errors ε_i propagate to the right, we cannot know just how much of an alternating effect on a higher difference can be traced to which tabulated point. If we suspect that only one tabulated point is in error, however, and further know that the function itself is smooth enough so that higher order derivatives (and hence differences) themselves have a constant sign, then we can use a common rule of thumb which states that the number of accurate digits in the tabulated values is approximately equal to the order of the highest difference which has predominantly nonalternating signs. In the case of the sines table, this is the fifth difference, so that the tabulated values are accurate to about five significant digits.

8.12 Newton Forward-Difference Formula

We now come to the Newton forward-difference formula, which is merely another way of writing the interpolating polynomial.

Let us consider first the first-degree formula and a simple justification for it. Suppose we take the unknown function $f(x)$, which is given to us only as a set of tabulated points, and expand it around the point x_0 in a Taylor series:

$$f(x) = f(x_0) + f'(x_0)(x - x_0) + f''(x_0)\frac{(x - x_0)^2}{2!} + \cdots.$$

Since $f(x)$ is given to us only as a table of values, we know $f(x_0)$ but not the derivatives at x_0. To simplify the situation, let us consider only the first two terms of the series and truncate the rest:

$$f(x) \approx f(x_0) + f'(x_0)(x - x_0).$$

Although we do not know the value of $f'(x_0)$ to use, we do know that there is a connection between derivatives and differences. Specifically, we have already shown that we can write

$$\Delta f(x) = f(x + h) - f(x) = hf'(\xi) \tag{8-12}$$

for some ξ between x and $x + h$. In the same way, we could write

$$\Delta f(x_0) = f(x_1) - f(x_0) = hf'(\xi)$$

$$f'(\xi) = \frac{\Delta f(x_0)}{h},$$

where, in this case, ξ lies somewhere between x_0 and x_1. If we assume that the interval between x_0 and x_1 is sufficiently small, then we could substitute $f'(\xi)$ into the Taylor series instead of $f'(x_0)$ with only a slight error, to get

$$f(x) \approx f(x_0) + \frac{\Delta f(x_0)}{h}(x - x_0).$$

The interesting property of this expression is that it happens to be the first-degree interpolating polynomial. That is, it is a first-degree polynomial which has the property that at $x = x_0$, the right-hand side is

$$f(x_0) + \frac{\Delta f(x_0)}{h}(x_0 - x_0) = f(x_0),$$

and at $x = x_1$, noting that $h = x_1 - x_0$, we have

$$f(x_0) + \frac{\Delta f(x_0)}{h}(x_1 - x_0) = f(x_0) + \frac{\Delta f(x_0)}{h}h$$

$$= f(x_0) + \Delta f(x_0)$$
$$= f(x_0) + [f(x_1) - f(x_0)]$$
$$= f(x_1).$$

Since the polynomial matches $f(x)$ exactly at x_0 and x_1, then it must be the interpolating polynomial, which is unique.

We now suspect that, had we started with three terms from the Taylor series and substituted the first two differences instead of the first two derivatives, we might have obtained the second-degree interpolating polynomial. This could have been done, except that it is rather difficult to perform. Rather, we shall approach the problem from a different direction.

Letting the quantity $\Delta f(x_0)/h$ be represented by the symbol a_1, we could write the first-degree interpolating polynomial $p_1(x)$ as

$$p_1(x) = f(x_0) + a_1(x - x_0).$$

The quantity $f(x_0)$ obviously matches $f(x)$ exactly at the lone tabulated point x_0 and could, in a way, be called the zeroeth-degree interpolating polynomial $p_0(x)$; hence we could write

$$p_1(x) = p_0(x) + a_1(x - x_0).$$

At $x = x_0$, only the first term contributes to the answer, while the second term is zero. At $x = x_1$, the $p_0(x)$ term provides a wrong answer (since it alone cannot match $f(x)$ at two points), but the second term corrects for the error and provides the correct result.

We now proceed to the second-degree interpolating polynomial. By analogy with the previous expression, we would expect the answer to be in the form

$$p_2(x) = p_1(x) + a_2(x - x_0)(x - x_1).$$

Here, we find that at $x = x_0$ and at $x = x_1$, the second term is zero and only the first term contributes to the answer. At $x = x_2$, on the other hand, the first term provides a slightly incorrect answer, but a_2 is chosen so that the second term provides the proper correction to make the overall result correct. Hence our problem is now reduced to the problem of finding the value of a_2. We can do this by writing out the expression for $p_2(x)$ in its long form as

$$p_2(x) = p_1(x) + a_2(x - x_0)(x - x_1)$$

$$= f(x_0) + \frac{\Delta f(x_0)}{h}(x - x_0) + a_2(x - x_0)(x - x_1).$$

Evaluating at $x = x_2$, we force $p_2(x_2)$ to equal $f(x_2)$. Also, we note that $(x_2 - x_0) = 2h$ and also that $(x_2 - x_1) = h$. Hence,

$$f(x_2) = f(x_0) + \frac{\Delta f(x_0)}{h}2h + a_2(2h)(h)$$

$$= f(x_0) + 2\,\Delta f(x_0) + 2h^2 a_2.$$

Solving for a_2, we have

$$a_2 = \frac{1}{2h^2}[f(x_2) - f(x_0) - 2\,\Delta f(x_0)]$$

$$= \frac{1}{2h^2}[f(x_2) - f(x_1) + f(x_1) - f(x_0) - 2\,\Delta f(x_0)]$$

$$= \frac{1}{2h^2}[\underbrace{\Delta f(x_1)}_{} + \underbrace{\Delta f(x_0)}_{} - 2\,\Delta f(x_0)]$$

$$= \frac{1}{2h^2}[\Delta f(x_1) - \Delta f(x_0)].$$

But the quantity in the brackets is $\Delta^2 f(x_0)$, so that

$$a_2 = \frac{\Delta^2 f(x_0)}{2h^2}$$

and therefore

$$p_2(x) = p_1(x) + \frac{\Delta^2 f(x_0)}{2h^2}(x - x_0)(x - x_1).$$

Repeating this entire process again, we could express p_3, p_4 and higher inter-polating polynomials in the same way; in general, we would obtain the re-cursive formula for any $p_i(x)$ as

$$p_i(x) = p_{i-1}(x) + \frac{\Delta^i f(x_0)}{i!h^i}(x - x_0)(x - x_1) \cdots (x - x_{i-1}).$$

What about the error in interpolating for some specific $\bar{x}$? Since the inter-polating polynomial is unique, the error formula derived earlier holds for the Newton forward-difference formula as well; thus, the error in using an nth-degree polynomial $p_n(x)$ to evaluate the function at $x = \bar{x}$ is

$$\varepsilon_n(\bar{x}) = f(\bar{x}) - p_n(\bar{x}) = f^{(n+1)}(\xi)\frac{(\bar{x} - x_0)(\bar{x} - x_1) \cdots (\bar{x} - x_n)}{(n + 1)!}.$$

If, on the other hand, we took Newton's forward-difference formula of degree $n + 1$ instead of degree n, we would have the expression

$$p_{n+1}(\bar{x}) = p_n(\bar{x}) + \frac{\Delta^{n+1} f(x_0)}{(n + 1)!h^{n+1}} (\bar{x} - x_0)(\bar{x} - x_1) \cdots (\bar{x} - x_n).$$

Looking at the second term on the right-hand side, we note that it is very similar to the expression for $\varepsilon_n(\bar{x})$, except that $f^{(n+1)}(\xi)$ is replaced by $\Delta^{n+1} f(x_0)/h^{n+1}$. As it turns out, these two quantities, while not equal, are of the same order of magnitude for most analytic functions (neglecting roundoff errors in calculating Δ^{n+1}). Hence the error is approximately equal to the first neglected term after taking $p_n(\bar{x})$.

At this point it is time for a quick comparison of the various ways of using the interpolating polynomial:

1. The "brute-force" approach, by using Eq. (8-2) and (8-3), seems simple but requires a very large amount of computation. It is therefore unpopular and potentially inaccurate as well.
2. The Lagrange method is theoretically useful in other applications, but also requires much computation. Every new interpolation, even with the same set of given data points, requires a completely new set of calculations.
3. Aitken's method is useful because it is easily extended to higher-degree interpolations as needed; on the other hand every new interpolation re-

quires a new computation, and so it is not too efficient if many interpolations need to be done.

4. Newton's forward-difference method is inefficient if only a few interpolations are needed; on the other hand, once a difference table is constructed, further interpolations using the same table are rather easy. Moreover, it is easily extended to higher-degree interpolations if more accuracy is needed.

8.13 Example 8-3. Newton's Forward-Difference Formula

For an example of Newton's formula for the interpolating polynomial, we take again the sample problem from Example 8-1, with a sines table tabulated for 0, 30, 60, and 90 degrees, and use p_0, p_1, p_2, and p_3 to interpolate to produce a table of sines for other angles.

Since zero subscripts are not permitted in FORTRAN the values of x_0 through x_3 are stored instead as X(1) through X(4); similarly, the values of the function $f(x_0)$ through $f(x_3)$ are stored as FX(1) through FX(4). The difference table for this problem is the following:

x_i	$f(x_i)$	$\Delta f(x_i)$	$\Delta^2 f(x_i)$	$\Delta^3 f(x_i)$
0	0.0			
		0.5		
30	0.5		-0.1339745962	
		0.3660254038		-0.0980762114
60	0.8660254038		-0.2320508076	
		0.1339745962		
90	1.0			

The calculation of the differences is done within the program in three DO loops. The outer DO loop, starting at the statement DO 50 NODIF = 1, 3 is repeated once for each order of differences. Within this loop, the DO loop ending at statement 30 calculates the differences and stores them in an array called TEMP, such that $\Delta f(x_0)$ is at TEMP(2) and $\Delta f(x_2)$ is at TEMP(4). The inner DO loop ending on statement 50 then moves these differences from the TEMP array back into FX(2) through FX(4).

When the overall DO loop is finished, then, the original values $f(x)$ in the FX array have been removed, and instead

$$f(x_0) \text{ is in FX(1)}$$
$$\Delta f(x_0) \text{ ,, ,, FX(2)}$$
$$\Delta^2 f(x_0) \text{ ,, ,, FX(3)}$$
$$\Delta^3 f(x_0) \text{ ,, ,, FX(4)}.$$

Note that this process would have to be changed if the tabulated values $f(x_i)$ were needed again later in some other calculation.

Following the calculation of the differences, the statements

DO 100 I = 10, 100, 10
XX = I − 10

start a repetition for 0 through 90 degrees. The FORTRAN SIN function is used to calculate a TRUE sine value, and the interpolating polynomials P0 through P3 are calculated from

$$p_0(x) = f(x_0)$$

$$p_1(x) = p_0(x) + \frac{\Delta f(x_0)}{h}(x - x_0)$$

$$p_2(x) = p_1(x) + \frac{\Delta^2 f(x_0)}{2!h^2}(x - x_0)(x - x_1)$$

$$p_3(x) = p_2(x) + \frac{\Delta^3 f(x_0)}{3!h^3}(x - x_0)(x - x_1)(x - x_2).$$

The complete program is the following:

```
        DIMENSION X(4), FX(4), TEMP (4)
        X(1) = 0.
        FX(1) = 0.
        X(2) = 30.
        FX(2) = 0.5
        X(3) = 60.
        FX(3) = 0.8660254038
        X(4) = 90.
        FX(4) = 1.0
        H = X(2) - X(1)
C       GENERATE DIFFERENCE TABLE
        DO 50 NODIF = 1, 3
        DO 30 NOF = NODIF, 3
 30     TEMP (NOF + 1) = FX(NOF + 1) - FX(NOF)
        DO 50 NOF = NODIF, 3
 50     FX(NOF + 1) = TEMP (NOF + 1)
C       PRINT DIFFERENCES
        WRITE (3, 55)
 55     FORMAT (10X, 'F(X0)', 6X, 'DF(X0)', 6X, 'D2F(X0)', 5X, 'D3F(X0)')
        WRITE (3, 60) FX
 60     FORMAT (5X, 4F12.8, ///)
C       PRINT TABLE HEADING
        WRITE (3, 65)
 65     FORMAT (26X, '---NEWTON FORWARD-DIFFERENCE POLYNOMIAL OF----')
        WRITE (3, 70)
 70     FORMAT ('    DEGREES', 7X, 'SINE', 6X, 'DEGREE 0', 4X, 'DEGREE 1',
     1    4X, 'DEGREE 2', 4X, 'DEGREE 3')
C       REPEAT FOR 0 TO 90 DEGREES
        DO 100 I = 10, 100, 10
        XX= I - 10
        TRUE = SIN (XX/ 57.2957795)
```

```
C     CALCULATE POLYNOMIALS P0, P1, P2, AND P3
      P0 = FX(1)
      P1 = P0 + FX(2) * (XX - X(1)) / H
      P2 = P1 + FX(3) * (XX-X(1)) * (XX- X(2)) / (2. * H**2)
      P3 = P2 + FX(4) * (XX - X(1)) * (XX - X(2)) * (XX - X(3)) /
    1       (3. * 2. * H**3)
  100 WRITE (3, 110)XX, TRUE, P0, P1, P2, P3
  110 FORMAT (F8.0, 3X, 5F12.6)
      CALL EXIT
      END
```

The program first prints out the following differences

F(X0)	DF(X0)	D2F(X0)	D3F(X0)
0.00000000	0.50000000	-0.13397459	-0.09807621

and then prints this interpolated table of sines:

		---NEWTON	FORWARD-DIFFERENCE	POLYNOMIAL OF---	
DEGREES	SINE	DEGREE 0	DEGREE 1	DEGREE 2	DEGREE 3
0.	0.000000	0.000000	0.000000	0.000000	0.000000
10.	0.173648	0.000000	0.166666	0.181552	0.175498
20.	0.342020	0.000000	0.333333	0.348219	0.343376
30.	0.500000	0.000000	0.500000	0.500000	0.500000
40.	0.642787	0.000000	0.666666	0.636894	0.641737
50.	0.766044	0.000000	0.833333	0.758903	0.764957
60.	0.866025	0.000000	1.000000	0.866025	0.866025
70.	0.939692	0.000000	1.166666	0.958261	0.941310
80.	0.984807	0.000000	1.333333	1.035612	0.987179
90.	1.000000	0.000000	1.500000	1.098076	1.000000

The third-degree polynomial naturally matches the true sine at the four tabulated points of 0, 30, 60, and 90 degrees; the second-degree polynomial gives the correct answer at 0, 30, and 60 degrees, but cannot provide the right answer at 90 degrees; the first-degree polynomial only provides the right answer at 0 and 30 degrees, while the zeroeth-degree polynomial only matches the sine at 0 degrees.

Another point of interest is the accuracy at, say, 10 degrees. This angle is between 0 and 30 degrees, and could therefore be interpolated using either the first-, second-, or third-degree polynomials. Note, however, that higher-degree polynomials generally give a more accurate answer.

8.14 Newton's Backward-Difference Formula

In most cases we cannot apply Newton's forward-difference formula directly and must make a few changes. Suppose, for example, that we are given a table of 100 tabulated points such as

$$
\begin{array}{ll}
x_0 & f(x_0) \\
x_1 & f(x_1) \\
x_2 & f(x_2) \\
\vdots & \\
x_{72} & f(x_{72}) \\
x_{73} & f(x_{73}) \\
x_{74} & f(x_{74}) \\
x_{75} & f(x_{75}) \\
\vdots & \\
x_{98} & f(x_{98}) \\
x_{99} & f(x_{99}),
\end{array}
$$

where the x_i are equally spaced. As a concrete example, suppose we seek an interpolated value $f(\bar{x})$ where $\bar{x}$ is between x_{73} and x_{74}, and that we decide, from an analysis of the acceptable error and the available calculation time, that a second-degree interpolation will be sufficient. We will need three tabulated points for a second-degree interpolation, and must now decide which three points to use.

To minimize the error, we should try to pick the three x_i closest to $\bar{x}$. If $\bar{x}$ is closer to x_{73} than to x_{74}, we would do well to choose x_{72}, x_{73}, and x_{74} as our three points (otherwise we would choose x_{73}, x_{74}, and x_{75}). When we now try to use the Newton forward-difference polynomial, we find that it is derived in terms of $f(x_0)$, $\Delta f(x_0)$, etc., not in terms of $f(x_{72})$, $\Delta f(x_{72})$ and so on. Hence some change of subscripts is required.

The easiest way to solve the problem is merely to renumber the points in the original table, to obtain a table such as

$$
\begin{array}{lll}
x_{-72} & f(x_{-72}) & [\text{originally } x_0 \text{ and } f(x_0)] \\
x_{-71} & f(x_{-71}) & \\
x_{-70} & f(x_{-70}) & \\
\vdots & & \\
x_0 & f(x_0) & [\text{originally } x_{72} \text{ and } f(x_{72})] \\
x_1 & f(x_1) & \\
x_2 & f(x_2) & \\
x_3 & f(x_3) & \\
\vdots & & \\
x_{26} & f(x_{26}) & \\
x_{27} & f(x_{27}) & [\text{originally } x_{99} \text{ and } f(x_{99})]
\end{array}
$$

We now have no difficulty using the forward-difference equation to inter-polate for $f(\bar{x})$, where $\bar{x}$ is now between x_1 and x_2. Moreover, since the pro-gram can be written to monitor its own error (by checking the size of the next term in the Newton formula), it could automatically take a few more terms to provide a higher-degree interpolation if needed. So far there is no difficulty.

Suppose on the other hand, that we need to interpolate at $\bar{x}$ where, this time, $\bar{x}$ is very close to x_{99} in the original table of values. We could of course, renumber the original table into

$$
\begin{array}{lll}
x_{-97} & f(x_{-97}) & [\text{originally } x_0 \text{ and } f(x_0)] \\
x_{-96} & f(x_{-96}) & \\
x_{-95} & f(x_{-95}) & \\
\quad\vdots & & \\
x_0 & f(x_0) & [\text{originally } x_{97} \text{ and } f(x_{97})] \\
x_1 & f(x_1) & [\text{originally } x_{98} \text{ and } f(x_{98})] \\
x_2 & f(x_2) & [\text{originally } x_{99} \text{ and } f(x_{99})],
\end{array}
$$

where $\bar{x}$ is now between x_1 and x_2. We again have no difficulty in obtaining a second-degree interpolating polynomial from Newton's forward-difference equation. If, however, we decide to expand it to a third-degree polynomial we are in trouble, because a third-degree interpolation needs values for x_3 and $f(x_3)$, which do not exist. Hence we would have to go back, again re-number the tabulated values, and start again. This is a difficult process, since it means that the program could not monitor its own error by taking another term in the Newton formula.

We therefore attack the problem from another direction—we take the entire tabulated set of values, and turn it "upside down", so to speak. That is, we renumber the points from the bottom up, in reverse order. The resulting table is then

$$
\begin{array}{lll}
x_{99} & f(x_{99}) & [\text{originally } x_0 \text{ and } f(x_0)] \\
x_{98} & f(x_{98}) & \\
x_{97} & f(x_{97}) & \\
\quad\vdots & & \\
x_2 & f(x_2) & [\text{originally } x_{97} \text{ and } f(x_{97})] \\
x_1 & f(x_1) & [\text{originally } x_{98} \text{ and } f(x_{98})] \\
x_0 & f(x_0) & [\text{originally } x_{99} \text{ and } f(x_{99})],
\end{array}
$$

where the desired point $\bar{x}$ is now located between x_0 and x_1, and we can take as high an interpolating polynomial as we wish, using the standard forward-difference formulas.

If we examine this process a bit more carefully, we see that we are taking differences which go backward in the renumbered table. Hence, this is often called the Newton *backward-difference* approach. Although we have approached the process by renumbering the table of points backward, another (and perhaps more common) way of getting the same result is to leave the points numbered as given, but to change the subscripts in Newton's forward-difference formula. The resulting formula is then called Newton's *backward-difference formula*. For all practical purposes, however, our way is just as good and certainly much easier to remember.

There are other difference interpolation methods, commonly called *central-difference* methods. They consist of taking the forward- and backward-difference formulas and combining them in such a way that both forward and backward differences are used. The aim is to get better accuracy for a given order of interpolation. The central-difference methods are more accurate and hence more popular; however, their derivation is beyond the scope of this introductory text.[5]

8.15 The Philosophy Behind Curve-Fitting

Since the process of curve-fitting can often require a bit of intuition, we will first look at the overall subject, and examine the possible approaches before we actually discuss specific methods.

As pointed out at the start of this chapter, both interpolation and curve-fitting start with a given table of values x_i and corresponding $f(x_i)$, such as:

x_i	$f(x_i)$
x_0	$f(x_0)$
x_1	$f(x_1)$
x_2	$f(x_2)$
$\vdots$	
x_n	$f(x_n)$

where we may wish to know $f(\bar{x})$ for some $\bar{x}$ different from the given x_i.

If we can assume that the tabulated values $f(x_i)$ are accurate (within any desired limits), then we find $f(\bar{x})$ by simply interpolating with some interpolating polynomial of degree 1 up to degree n, depending on the desired accuracy of the answer.

---

[5] There are many good references, such as S. D. Conte, *Elementary Numerical Analysis*, McGraw-Hill, New York, 1965, Chapter 3.

Often, however, the tabulated $f(x_i)$ are known to be only approximate. They could, for example, arise from a set of measurements on some physical process, so that their accuracy depends on the accuracy of the measuring instruments, the skill of the person taking and recording the measurements, and many other factors. We can clearly assume that each measured $f(x_i)$ is in error by some unknown amount ε_i.

Now, it may happen that all of these errors ε_i are of the same sign. This could easily be caused by some systematic error such as a defective or un-calibrated measuring instrument which consistently reads too high or too low. In such a case each of the measured $f(x_i)$ could be high or low by some relatively constant amount.

More often, however, the ε_i are of varying signs, caused by random and unpredictable errors which might at some times be positive, other times be

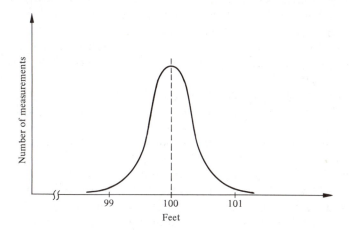

Figure 8-6.   Gaussian distribution of random errors.

negative. Random and unpredictable errors have various interesting characteristics studied by statisticians in their daily work. One of these is the fact that, although we cannot predict any one particular error by itself, we can make some predictions about large numbers of errors.

Let us take a particular example. Suppose we must measure the length of a plot of land which, unknown to us, measures exactly 100 feet long. Unfortunately, we have only a 12-inch ruler with which to perform the measurement. We might measure the plot of land many times, and each time obtain a different answer; all of these answers would be around 100 feet, but they might differ by several inches from the true value.

If we took many measurements (in the thousands), we might decide to plot a graph of the measured values and the number of times each of those values was obtained, as in Figure 8-6. If enough measurements were taken, the resulting curve would be the familiar "bell curve" as shown, which gives the distribution of the errors around the true value of 100 feet—this type of error distribution is called a *Gaussian* or *normal distribution*.

Looking at Figure 8-6, we see that most of the measurements are grouped very close to 100 feet, and relatively few of the measurements are below 99 or above 101 feet. This illustrates two important facts about random errors; the average value of a large number of measurements is close to the correct value (or alternatively, we may say that the average error is zero if we consider some errors positive and others negative), and moreover, small errors are more likely than large errors although, in a large number of measurements, there may still be a few very large errors.

This suggests that random errors can somehow be averaged out of a table of data if we have enough measurements. One rather obvious approach is to take many measurements of the same quantity; in terms of the table of values given to us, this implies that we have several (slightly different) measured values of $f(x_i)$ for each x_i, which we may average to obtain a hopefully more accurate value. In the general case, however, this may not be feasible, either because it is not possible or practical to repeat the same measurement many times, or else because whoever took the measurements originally was simply not that obliging.

This is where curve-fitting comes in. It is essentially an averaging process which uses single measurements of many different quantities instead of many measurements of a single quantity. It tries to average out errors by assuming that behind the somewhat erratic measured values is some orderly process which basically follows some relatively simple equation.

Figures 8-1 and 8-2 have already illustrated a particular example, where we assumed that, in testing the performance of an automobile, we measured its speed at various times during acceleration from a full stop. As part of this experiment, we may have obtained the following table of values:

Time in Seconds (x_i)	Speed in MPH $f(x_i)$
1	8
2	22
3	29
4	32
5	42
6	42
7	47
8	54
9	56
10	58

Each of these sets of values could define a point as in Figure 8-2. Since we

have ten given values, we could fit these points with a ninth-degree interpolating polynomial which would pass through all ten points as in Figure
8-2. This interpolating polynomial could then be used if we needed the speed
at any other time.

On the other hand, we know that the motion of a car is usually not that
erratic, and so we suspect that a smoothed curve such as the one in Figure
8-1 probably describes the motion of the automobile better, even though it
may not pass through any of the tabulated points. We are essentially assuming that the motion of the car follows some rather simple equation which, if
graphed, would be smooth.

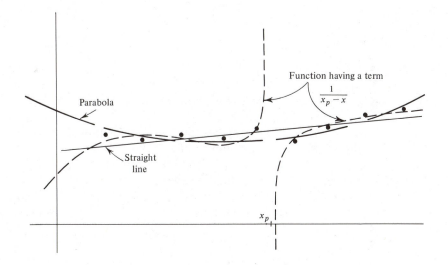

Figure 8-7.   Three "smooth" curves fitting a given set of tabulated data.

At this point we must make a decision. Since it is possible to fit a number
of "smooth" curves to the same set of data, just how "smooth" should the
curve be, and how do we define a "good fit"? Here a bit of judgement is
needed.

Figure 8-7 shows a typical set of data such as might be obtained from some
experimental source, as well as three different "smooth" curves which might
fit that data. The solid points in the figure show the tabulated data. The
smoothest of the three curves is the solid line, with the dashed parabola
being a close second. Least smooth of all is the third curve which, having a
term $1/(x_p - x)$, becomes infinite when $x = x_p$. Which of these three will
we accept as the "best fit"?

An even worse case is shown in Figure 8-8. Suppose we are given the points
shown. Looking at them, we decide that the curve could be very smooth if
we merely assumed that the point below the x axis is simply a gross error,

and that we will disregard it, and draw the smooth curve shown. Certainly tempting.

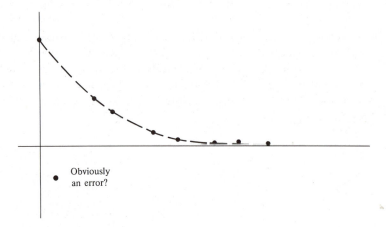

Figure 8-8.   Can we assume that the one point which doesn't fit is simply an error?

Figure 8-9 shows the real source of these points, an oscillating curve which decays with time. We have here stumbled onto a possible pitfall of curve-fitting; we may consider ourselves lucky to have found a nice curve which fits a set of observed data when, in fact, we are completely wrong.

Hence, before starting an attempt to fit a set of data points with a smooth curve, we must first decide just what kind of a curve we want—a straight

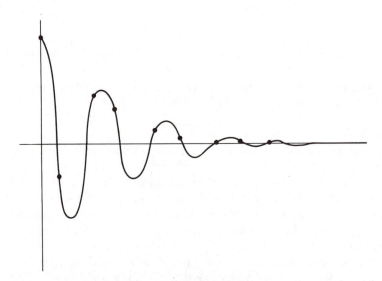

Figure 8-9.   The actual curve for the points shown in Figure 8-8.

line? a parabola? a cubic? a higher-degree polynomial? a trigonometric function? an exponential? To make an intelligent decision, we should attempt to first find out where the original data points came from, since we can often guess the general shape of the curve from this information.

Very often a set of observed data points can be fitted quite well with a polynomial. Let us assume that we have been given $n + 1$ tabulated values $x_0, x_1, \ldots, x_n$ and their corresponding values $f(x_0), f(x_1), \ldots, f(x_n)$, and that we decide to approximate these points with a polynomial of degree m.

There are now three possibilities, depending on whether m equals n, is greater than n, or is less than n. Let us examine them in turn.

1. Suppose $m = n$. Then we have $n + 1$ points which we wish to approximate with a polynomial of degree n. As we have seen in the earlier part of this chapter, it is possible to develop an interpolating polynomial $p_n(x)$ which exactly passes through any $n + 1$ distinct points, previously assigned. But as we have already indicated, this may not be desirable if these $n + 1$ points are inaccurate, since the result may be a curve which is highly erratic.
2. Suppose $m > n$. A polynomial of degree m could be an excellent interpolating polynomial for a set of $m + 1$ distinct points. We could therefore take the original $n + 1$ given points, add on any arbitrary set of $(m - n)$ other points, and obtain an mth-degree polynomial which would pass through all of these points. Such an mth-degree polynomial would have more nonzero derivatives than the nth-degree polynomial from case 1 above, and would be even less smooth. Thus we want to be sure not to use a fitted curve of too high a degree.
3. Suppose $m < n$. Such a polynomial curve could be made to pass through any combination of $m + 1$ points out of the original $n + 1$ points, but in general would have to bypass the remaining $(n - m)$ points (unless they had happened to be chosen in some special way). Alternatively, this polynomial could be arranged to pass *near* to the tabulated points, but not necessarily *through* any of them. The resulting curve, on the other hand, would have fewer nonzero derivatives than an nth-degree interpolating polynomial, and would therefore be smoother. This latter case is the desirable one.

Let us therefore assume that we have $n + 1$ points which we are going to approximate with an mth-degree polynomial, $m < n$:

$$p_m(x) = a_m x^m + a_{m-1} x^{m-1} + \cdots + a_2 x^2 + a_1 x + a_0. \qquad (8\text{-}13)$$

At each of the $n + 1$ tabulated points, the resulting $p_m(x_i)$ will be approximately equal to the tabulated $f(x_i)$, with an error which we shall call δ_i; we might call this the *deviation* of the fitted curve from the tabulated point.

We could then write the following $n + 1$ equations, one for each tabulated point:

$$p_m(x_0) - f(x_0) = \delta_0$$
$$p_m(x_1) - f(x_1) = \delta_1$$
$$\vdots$$
$$p_m(x_n) - f(x_n) = \delta_n,$$

or, in general,

$$p_m(x_i) - f(x_i) = \delta_i \qquad \text{for } i = 0, 1, 2, \ldots, n.$$

It is now time to decide what we mean by a "best fit". Can we make all of the δ_i equal to zero? Generally not, since this would require the fitted curve to actually pass through all $n + 1$ points, and this is not possible if $m < n$.

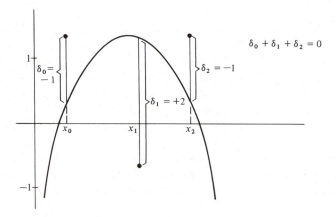

Figure 8-10. Bad fit where the sum of the δ_i is zero.

Can we make the sum of all the δ_i equal to zero? We can, quite easily, but this is not desirable, as we can see from Figure 8-10, where a concave-downward parabola has been fitted to three points when obviously a concave-upward parabola could have been used even better.

But we are now on the right track. Can we make the sum of the absolute values of the δ_i equal to zero? No, since then each of the δ_i would in turn have to be zero. On the other hand, we can try to minimize this sum of absolute values, which can be written as the sum

$$\sum_{i=0}^{n} |\delta_i| = |\delta_0| + |\delta_1| + |\delta_2| + \cdots + |\delta_n|.$$

It is possible to minimize this sum, but the use of absolute values makes the calculations somewhat awkward. To remove the absolute values and yet preserve the plus signs (to keep the δ_i from canceling each other out, as in Figure 8-10), we may decide instead to minimize the sum of the *squares* as in

$$\sum_{i=0}^{n} (\delta_i)^2 = \delta_0^2 + \delta_1^2 + \delta_2^2 + \cdots + \delta_n^2.$$

The resulting expression is not only easier to minimize analytically, but also has the advantage that it discriminates against large deviations δ_i. This is in agreement with Figure 8-6, where we see that large errors in measurement are much less likely than small errors; hence large deviations between the tabulated points and the smoothed curve should also be less likely. Minimizing the sum of the squares will keep us from having large deviations where possible.

This method of minimizing the sum of the squares is called *least-squares* curve-fitting, and is the most common technique.

8.16 Least-Squares Curve-Fitting

In the last section we assumed that we were given a table of $n + 1$ points which were to be fitted with a curve, having an equation of the form

$$p_m(x) = a_m x^m + a_{m-1} x^{m-1} + \cdots + a_2 x^2 + a_1 x + a_0. \quad (8\text{-}13)$$

It may happen, however, that an mth-degree polynomial, such as Eq. (8-13), may not represent the best fit. For example, the original table of points may have come from some oscillating or decaying process, in which case a sine or cosine wave, or an exponential, might have been a more reasonable equation to use than a plain polynomial.

Let us therefore relax the requirements on $p_m(x)$ somewhat by permitting forms other than just polynomials. In general, let us permit the fitted curve to have the form

$$p_m(x) = a_m g_m(x) + a_{m-1} g_{m-1}(x) + \cdots + a_1 g_1(x) + a_0 g_0(x), \quad (8\text{-}14)$$

where the functions $g_m(x), \ldots, g_0(x)$ are assumed to be some known functions of x.

We can easily return to an mth-degree polynomial as a special case by merely letting

$$g_m(x) = x^m$$
$$g_{m-1}(x) = x^{m-1}$$
$$\vdots$$
$$g_1(x) = x^1$$
$$g_0(x) = 1.$$

To make sure that the coefficients a_m, a_{m-1}, etc. are unique we must choose the functions $g_i(x)$ in such a way that no $g_i(x)$ can be expressed as a *linear* combination of any of the other functions $g_i(x)$; we say that the functions $g_i(x)$ must all be linearly independent. To prove this statement, suppose that it were *not* true and that some $g_i(x)$ could be written as a linear combination of some $g_j(x)$ and $g_k(x)$ like this:

$$g_i(x) = b_j g_j(x) + b_k g_k(x), \tag{8-15}$$

for some constants b_j and b_k. Then we could simply eliminate the $g_i(x)$ term from Eq. (8-14) altogether, and substitute the right-hand-side of Eq. (8-15) instead. This is equivalent to adding b_j and b_k to the existing coefficients a_j and a_k. Alternatively, we could eliminate any part of $g_i(x)$ from the expression for $p_m(x)$, Eq. (8-14), by adjusting a_j and a_k to compensate. In fact, there would be an infinite number of ways of writing the same function $p_m(x)$, and the values for the coefficients a_i would not be unique.

We can now see that the functions

$$g_m(x) = x^m$$
$$g_{m-1}(x) = x^{m-1}$$
$$\vdots$$
$$g_1(x) = x^1$$
$$g_0(x) = 1$$

are linearly independent, since they cannot be obtained from each other by linear combinations. But we could also let the functions $g_i(x)$ be sines, cosines, exponentials, or many other functions.

So let us now return to the basic problem. We have a table of $n + 1$ points, and seek an equation of the form

$$p_m(x) = a_m g_m(x) + a_{m-1} g_{m-1}(x) + \cdots + a_1 g_1(x) + a_0 g_0(x) \tag{8-14}$$

to match the $n + 1$ given points in such a way that, if we formed a set of deviations, one for each tabulated x_i,

$$\delta_0 = p_m(x_0) - f(x_0)$$
$$\delta_1 = p_m(x_1) - f(x_1)$$
$$\delta_2 = p_m(x_2) - f(x_2)$$
$$\vdots$$
$$\delta_n = p_m(x_n) - f(x_n),$$

or, in general terms, the set of deviations

$$\delta_i = p_m(x_i) - f(x_i) \qquad \text{for } i = 0, 1, 2, \ldots, n \tag{8-16}$$

then the sum of the squares of these $n + 1$ deviations should be a minimum:

$$\sum_{i=0}^{n} (\delta_i)^2 = \text{a minimum}. \tag{8-17}$$

The summation in Eq. (8-17) can also be written as

$$\sum_{i=0}^{n} (\delta_i)^2 = \sum_{i=0}^{n} [p_m(x_i) - f(x_i)]^2 \tag{8-18}$$

$$\sum_{i=0}^{n} (\delta_i)^2 = \sum_{i=0}^{n} [a_m g_m(x_i) + \cdots + a_0 g_0(x_i) - f(x_i)]^2 \tag{8-19}$$

In essence, we are trying to minimize the right-hand side of Eq. (8-19). Since the x_i and $f(x_i)$ are given in the original table, and the $g_i(x)$ are known functions which can easily be evaluated at the given x_i, the only unknowns in (8-19) are the coefficients a_i. Hence we can look on the summation as a function of the $m + 1$ coefficients a_i with all other terms being known constants.

In elementary calculus whenever we wish to minimize a function we set its first derivative equal to zero and solve for the unknown variable. In our present case the function to be minimized, Eq. (8-19), is actually a function of the $m + 1$ unknown variables a_i. Let us therefore find the partial derivative of the summation with respect to *each* of the $m + 1$ variables a_i, and set *each* of these derivatives equal to zero:

$$\frac{\partial}{\partial a_0} \sum_{i=0}^{n} (\delta_i)^2 = 0$$

$$\frac{\partial}{\partial a_1} \sum_{i=0}^{n} (\delta_i)^2 = 0$$

$$\vdots$$

$$\frac{\partial}{\partial a_m} \sum_{i=0}^{n} (\delta_i)^2 = 0,$$

or, in general terms, we can write

$$\frac{\partial}{\partial a_j} \sum_{i=0}^{n} (\delta_i)^2 = 0 \qquad \text{for } j = 0, 1, 2, \ldots, m. \tag{8-20}$$

The series has a finite number of terms, and so the derivative of a sum is equal to the sum of the derivatives of the terms. Thus we can write Eq. (8-20) as

$$\frac{\partial}{\partial a_j} \sum_{i=0}^{n} (\delta_i)^2 = \sum_{i=0}^{n} \frac{\partial}{\partial a_j} (\delta_i)^2$$

$$= \sum_{i=0}^{n} 2\delta_i \frac{\partial \delta_i}{\partial a^j} = 0$$

or,

$$\sum_{i=0}^{n} \delta_i \frac{\partial \delta_i}{\partial a_j} = 0 \qquad \text{for } j = 0, 1, 2, \ldots, m. \tag{8-21}$$

The differentiation of any δ_i with respect to a_j is easy when we note that

$$\delta_i = p_m(x_i) - f(x_i)$$
$$= a_m g_m(x_i) + a_{m-1} g_{m-1}(x_i) + \cdots + a_j g_j(x_i) + \cdots$$
$$+ a_1 g_1(x_i) + a_0 g_0(x_i) - f(x_i). \tag{8-22}$$

When we differentiate with respect to a_j, we assume that every other term in (8-22) is constant except the term containing a_j, so that all other terms drop out upon differentiating, and

$$\frac{\partial \delta_i}{\partial a_j} = 0 + 0 + \cdots + \frac{\partial}{\partial a_j} a_j g_j(x_i) + 0 + \cdots + 0$$

$$\frac{\partial \delta_i}{\partial a_j} = g_j(x_i). \tag{8-23}$$

Now we substitute Eq. (8-23) into Eq. (8-21), and obtain the $m + 1$ equations

$$\sum_{i=0}^{n} \delta_i g_j(x_i) = 0 \qquad \text{for } j = 0, 1, 2, \ldots, m. \tag{8-24}$$

Substituting Eq. (8-16) into (8-24), we have

$$\sum_{i=0}^{n} [p_m(x_i) - f(x_i)] g_j(x_i) = 0 \qquad \text{for } j = 0, 1, 2, \ldots, m, \tag{8-25}$$

and then substituting Eq. (8-14) into (8-25) we have

$$\sum_{i=0}^{n} [a_m g_m(x_i) + \cdots + a_1 g_1(x_i) + a_0 g_0(x_i) - f(x_i)] g_j(x_i) = 0$$

$$\text{for } j = 0, 1, 2, \ldots, m. \quad (8\text{-}26)$$

Multiplying through on the left-hand side

$$\sum_{i=0}^{n} [a_m g_m(x_i) g_j(x_i) + \cdots + a_1 g_1(x_i) g_j(x_i)$$

$$+ a_0 g_0(x_i) g_j(x_i) - f(x_i) g_j(x_i)] = 0 \qquad \text{for } j = 0, 1, 2, \ldots, m.$$

We now break up the left-hand side into separate sums:

$$\sum_{i=0}^{n} a_m g_m(x_i) g_j(x_i) + \cdots + \sum_{i=0}^{n} a_1 g_1(x_i) g_j(x_i)$$

$$+ \sum_{i=0}^{n} a_0 g_0(x_i) g_j(x_i) - \sum_{i=0}^{n} f(x_i) g_j(x_i) = 0$$

$$\text{for } j = 0, 1, 2, \ldots, m.$$

Next, we factor the a's out of each sum:

$$a_m \sum_{i=0}^{n} g_m(x_i) g_j(x_i) + \cdots + a_1 \sum_{i=0}^{n} g_1(x_i) g_j(x_i)$$

$$+ a_0 \sum_{i=0}^{n} g_0(x_i) g_j(x_i) = \sum_{i=0}^{n} f(x_i) g_j(x_i)$$

$$\text{for } j = 0, 1, 2, \ldots, m. \quad (8\text{-}27)$$

Eq. (8-27) finally provides a useful result. Each of the summations can be evaluated since the g_i are known functions and the x_i are tabulated values. Thus only the a_i are unknown. For $j = 0, 1, 2$, etc., up to $j = m$, we now have $m + 1$ equations in the $m + 1$ unknowns a_m, a_{m-1}, through a_1 and a_0. These are linear equations which can be solved using the methods of Chapter 5.

To simplify the process a bit and make it more suitable for computer solution, we note that the coefficient of a_k in the jth equation is

$$\alpha_{kj} = \sum_{i=0}^{n} g_k(x_i) g_j(x_i) \qquad \begin{array}{l} \text{for } k = 0, 1, 2, \ldots, m \\ j = 0, 1, 2, \ldots, m. \end{array} \quad (8\text{-}28)$$

Since j and k in Eq. (8-28) can be interchanged, we have also the result that

$\alpha_{kj} = \alpha_{jk}$ which reduces the work required to compute these coefficients by about half.

We now have a workable scheme for finding the least-squares fit. Using Eq. (8-27) we solve for the $m + 1$ unknown a's, which exist and are unique if we choose the $g(x)$ functions properly, and then substitute the resulting values into Eq. (8-14) to give us the desired expression for $p_m(x)$. We shall now illustrate this procedure with an example.

8.17 Example 8-4. Least-Squares Curve-Fitting

For our final example, suppose that a heavy object is dropped from a height of 1100 feet. Letting x stand for the time, in seconds, after the object is dropped, we could solve for the equation of motion using elementary physics as $f(x) = 1100 - 16x^2$. Using this equation we could then calculate the height above the ground during the first four seconds:

Time *(seconds)*	Height *(feet)*
0	1100
1	1084
2	1036
3	956
4	844

Rather than employ this analytical method, suppose we actually perform this experiment, and measure the height to the nearest ten feet. We will then have the following table of measured values:

x_i	$f(x_i)$
$x_0 = 0$	$f(x_0) = 1100$
$x_1 = 1$	$f(x_1) = 1080$
$x_2 = 2$	$f(x_2) = 1040$
$x_3 = 3$	$f(x_3) = 960$
$x_4 = 4$	$f(x_4) = 840$

We would now like to fit a least-squares curve to the measured data. Knowing where the data originated, we suspect that the best fit will be obtained with a parabola, and so we look for a second-degree polynomial $p_2(x)$ as a solution. To achieve this, we let

$$g_2(x) = x^2$$
$$g_1(x) = x$$
$$g_0(x) = 1,$$

so that from Eq. (8-14) we seek a curve of the form

$$p_2(x) = a_2 x^2 + a_1 x + a_0,$$

as the solution. Hence $m = 2$, and since we have five given tabulated points,

$$n + 1 = 5$$
$$n = 4.$$

Thus we will have $m + 1 = 3$ equations to solve for the three unknowns a_2, a_1, and a_0.

From Eq. (8-28) the coefficients α_{kj} of a_k in the jth equation are found as follows:

$$\alpha_{00} = \sum_{i=0}^{4} g_0(x_i)g_0(x_i) = 1 + 1 + 1 + 1 + 1 = 5,$$

since $g_0(x) = 1$ for any x. Similarly, we find

$$\alpha_{01} = \alpha_{10} = \sum_{i=0}^{4} g_1(x_i)g_0(x_i) = x_0 + x_1 + x_2 + x_3 + x_4$$
$$= 0 + 1 + 2 + 3 + 4$$
$$= 10;$$

$$\alpha_{02} = \alpha_{20} = \sum_{i=0}^{4} g_2(x_i)g_0(x_i) = \sum_{i=0}^{4} x_i^2 \cdot 1$$
$$= x_0^2 + x_1^2 + x_2^2 + x_3^2 + x_4^2$$
$$= 0^2 + 1^2 + 2^2 + 3^2 + 4^2$$
$$= 30;$$

$$\alpha_{11} = \sum_{i=0}^{4} g_1(x_i)g_1(x_i) = \sum_{i=0}^{4} x_i \cdot x_i = \sum_{i=0}^{4} x_i^2$$
$$= x_0^2 + x_1^2 + x_2^2 + x_3^2 + x_4^2$$
$$= 30;$$

$$\alpha_{12} = \alpha_{21} = \sum_{i=0}^{4} g_2(x_i)g_1(x_i) = \sum_{i=0}^{4} x_i^2 \cdot x_i = \sum_{i=0}^{4} x_i^3$$
$$= 0^3 + 1^3 + 2^3 + 3^3 + 4^3$$
$$= 100;$$

$$\alpha_{22} = \sum_{i=0}^{4} g_2(x_i)g_2(x_i) = \sum_{i=0}^{4} x_i^2 \cdot x_i^2 = \sum_{i=0}^{4} x_i^4$$
$$= 0^4 + 1^4 + 2^4 + 3^4 + 4^4$$
$$= 354;$$

Using the α's, we have the three equations

$$\alpha_{20}a_2 + \alpha_{10}a_1 + \alpha_{00}a_0 = \sum_{i=0}^{n} f(x_i)g_0(x_i)$$

$$\alpha_{21}a_2 + \alpha_{11}a_1 + \alpha_{01}a_0 = \sum_{i=0}^{n} f(x_i)g_1(x_i)$$

$$\alpha_{22}a_2 + \alpha_{12}a_1 + \alpha_{02}a_0 = \sum_{i=0}^{n} f(x_i)g_2(x_i).$$

Evaluating the summations on the right, and substituting the values for the α's, we finally obtain the following three equations in three unknowns:

$$
\begin{aligned}
30a_2 + 10a_1 + 5a_0 &= 5020 \\
100a_2 + 30a_1 + 10a_0 &= 9400 \\
354a_2 + 100a_1 + 30a_0 &= 27320.
\end{aligned}
$$

Solving these equations for the three unknowns by the methods of Chapter 5 gives us the results:

$$
\begin{aligned}
a_2 &= -17.14285714 \\
a_1 &= 4.573428573 \\
a_0 &= 1097.714286,
\end{aligned}
$$

so that the polynomial fitted to the data has the form (accurate to ten significant digits)

$$p_2(x) = -17.14285714x^2 + 4.573428573x + 1097.714286. \quad (8\text{-}29)$$

It is interesting to compare this equation with the original data:

x_i	Accurate Value if Determined from the Correct Equation $f(x) = 1100 - 16x^2$	Inaccurate Value as "Measured" to the Nearest 10 feet	Value Obtained from the Least-Squares Fitted Curve [Eq. (8-29)]
0	1100	1100	1097.7
1	1084	1080	1085.1
2	1036	1040	1038.3
3	956	960	957.1
4	844	840	841.7

As you can see, the least-squares curve fitted to the slightly inaccurate starting data is, first of all, a reasonable equation for the data, and moreover tends to balance out the randomly introduced errors. At four of the five given points the least-squares curve is closer to the true value than the original tabulated point was; only at one point, where the given tabulated value happened to be exact, is the least-squares curve worse than the measured value. Thus we have started with five given points, and have developed an equation which can be used to calculate other values *in the range of the original table*. We are not justified in thinking, of course, that the least-squares curve will be a good approximation at, say, $x = 8$.

Lest the author be accused of having done *all* the work for you, the writing of the computer program for this, the last sample problem in the text, is left to you as an exercise.

8.18 Problems

1. Given the following three points:

$$x_0 \quad f(x_0)$$
$$x_1 \quad f(x_1)$$
$$x_2 \quad f(x_2).$$

   a. Find the coefficients in the second-degree interpolating polynomial $p_2(x) = a_2 x^2 + a_1 x + a_0$ in terms of the x_i and $f(x_i)$ given.
   b. Express the interpolating polynomial of second degree in the Lagrange form.
   c. Show that the polynomials in *a* and *b* above are the same.

2. For the following four points, express the third-degree interpolating polynomial in its:

   a. $p_3(x) = a_3 x^3 + a_2 x^2 + a_1 x + a_0$ form.
   b. Lagrange form.
   c. Aitken's form.
   d. Newton's forward difference form.
   e. Newton's backward difference form.

i	x_i	$f(x_i)$
0	0	10
1	2	7
2	4	0
3	6	−11

3. Repeat problem 2 for the following table; what can you say about $f(x)$?

i	x_i	$f(x_i)$
0	0	10
1	1	8
2	2	2
3	3	-8

4. A frequent use of interpolation, given a function $y = f(x)$, is to find x from some given value of y; this process is called *inverse interpolation*, and is performed by simply interchanging the roles of x and y.

   a. Using the following four tabulated values, write the third-degree Lagrange interpolating polynomial:

y	$x = \sin^{-1} y$
0.0	0 degrees
0.5	30
0.8660254038	60
1.0	90

   b. Find the angle whose sine is 0.6 and compare your answer with a book of tables.

5. Discuss the use of inverse interpolation to find the roots of equations; compare with the method of false position (see Chapter 3).

6. Suppose we need the sine of 6.5 degrees to an accuracy of ± 0.0001; having a table of sines tabulated at 10-degree intervals, how high a degree of interpolation do we need for the required accuracy?

7. One of the early mechanical "computers" was Charles Babbage's Difference Engine, which was to calculate the values of functions by means of tables of differences, using some rather complicated gearing schemes. Show how this arrangement could be used to calculate $f(7)$ from the following data:

$$f(5) = 700,$$
$$f(5.5) = 616,$$
$$f(6) = 524,$$
$$f(6.5) = 424.$$

8. The function of problem 7 represents a polynomial; what degree is it?

9. Given the data plotted in Figure 8-1, find a least-squares fit of the form $v = \sqrt{t}$ and plot the resulting function.

10. Separate the problem discussed in Example 8-4 into appropriate portions, and write a FORTRAN program which will perform all required calculations from start to finish.

11. In Example 1-2 we discussed the problems of differentiating a function by use of the relation

$$\frac{d}{dx} f(x) = \lim_{\Delta x \to 0} \frac{f(x + \Delta x) - f(x)}{\Delta x}$$

and specifically showed that differentiating $\ln x$ and evaluating at $x = 3$ gave a rather large error. Let us now attack the problem another way, using interpolation. Write and run a FORTRAN program which will do the following to evaluate $d/dx(\ln x)$ at some particular value $x = \bar{x}$.

a. Find $\ln \bar{x}$, $\ln (\bar{x} - 0.1)$ and $\ln (\bar{x} + 0.1)$. If $\bar{x}$ itself is 0.1 or less, use a smaller increment to avoid problems.

b. Find the coefficients a_2, a_1, and a_0 in the interpolating polynomial $p_2(x) = a_2 x^2 + a_1 x + a_0$ which fits the $\ln x$ function at the three points given.

c. Differentiating the interpolating polynomial, find the desired value from

$$p_2'(\bar{x}) = 2a_2\bar{x} + a_1.$$

d. Using this procedure, find the derivative at $\bar{x} = 3$, and compare with the correct value of $\frac{1}{3}$.

e. Repeat, using a fourth-degree interpolating polynomial.

Index